MODULAR COURSES IN TECHNOLOGY

AERONAUTICS

Ray Page

Steven Bloor

John Henson

OLIVER & BOYD

in association with Trent International Centre for School Technology

PROJECT TEAM

Director
Dr Ray Page

Co-ordinators
Roy Pickup
John Poole

Jeffrey Hall
Dr Duncan Harris
John Hucker
Michael Ive
Peter Patient

Oliver & Boyd
Robert Stevenson House
1–3 Baxter's Place
Leith Walk
Edinburgh EH1 3BB

A Division of Longman Group UK Ltd

ISBN 0 05 003787 0
First published 1988

© SCDC Publications 1988
All rights reserved; no part of this publication may be reproduced, stored in a retrieval system, or transmitted in any form or by any means, electronic, mechanical, photocopying, recording or otherwise, without the prior written permission of the Publishers.

Set in 11/13pt Times Roman Linotron 202

Produced by Longman Group (FE) Ltd
Printed in Hong Kong

Contents

    Preface 5
1. The Story of Flight 7
2. Balloons and Kites 13
3. Gliders and the Wright Brothers 22
4. Aircraft Structures 30
5. The Principles of Powered Flight 40
6. Power Units 55
7. Stability and Control 72
8. The Helicopter 84
9. Aircraft Instruments 87
10. High Speed Flight and Air Traffic Control 91
11. Aircraft in the Service of Man 100
12. Design Characteristics 120
    Index 124

Acknowledgments

For permission to reproduce the photographs in this book, the authors and publishers would like to thank the following:

Trustees of the Science Museum, London (Figs. 1.3, 1.4, 1.5, 1.6, 3.1, 3.2, 3.3, 3.4, 11.2, 11.4, 11.5, 11.6, 11.9, 11.10 and 11.15); Imperial War Museum (Fig. 2.1); The Photo Source (Figs. 2.3, 2.4 and 3.9); Royal Air Force Museum (Figs. 4.6, 11.3, 11.7, 11.8, 11.12a and b, 11.13, 11.14, 11.16, 11.17, 11.20, 11.21 and 11.23); The British Hovercraft Corporation Ltd (Fig. 5.6a and b); British Airways (Figs. 7.11c, 11.18 and 11.26); Flight/Quadrant Picture Library (Figs. 7.12b, 11.1, 11.11 and 11.25); NASA (Figs. 9.5 and 11.28); Civil Aviation Authority (Fig. 10.10); Royal Air Force Inspectorate of Recruiting – crown copyright (Fig. 11.19); Military Aircraft Photographs (Figs. 11.22 and 11.24); NASA/Science Photo Library (Fig. 11.27).

Thanks are also due to Loganair for their help with Fig. 9.2 and to British Airways for Fig. 12.1.

Preface

This book has been written to give a view of flight and aircraft through the eyes of the designers and engineers who make the aircraft, the engineers and technicians who maintain them, and the aircrew who operate them.

The aircraft industry is of great importance to this country as aircraft provide affordable transport over long distances. This has done much to shape and improve the world over the past two decades. The industry has also generated wealth from the export of aircraft, and has created employment for thousands of people.

Air transport also has its disadvantages. Aircraft use enormous quantities of fuel; they can be very noisy and they can cause air pollution. There is, however, continued research to improve energy consumption, and there are tight controls that govern noise and air pollution levels. Many of these disadvantages are thus kept to a minimum.

As always, advantages have to be weighed against disadvantages. The writers of this book believe that there is a net benefit in the use of aircraft. They see the aircraft industry as one of our 'high technology' industries, an industry with a great future.

Having used the word technology, we should make sure that we are agreed about what this word means. Technology is the activity of bringing together the knowledge and skills required to solve problems. Usually this is achieved by producing a machine, or a piece of equipment that will let people overcome the problem. Technologists need to have practical skills, an understanding of basic scientific principles, and an appreciation of the total costs. Total costs means costs both to the user and to the people affected by the use of the machine or equipment.

Technology depends upon scientific design and practical skills. All of these have to be mastered sufficiently for the result to be satisfactory. The study of aerodynamics by itself is a scientific discipline. The application of this study to solve problems and how these solutions affect our society are examples of technological activity. This demonstrates that the frontiers of knowledge can progress – almost explosively – after the production of even a very few new ideas. Such progress can only be achieved by large groups of people working together and pooling their skills and ideas. The scientific principles can then become part of a routine.

Note The ■ sign has been used in the text to signify the basic course material. The ☐ sign indicates the more advanced sections.

1 The Story of Flight

■ **Kites**

The story of flight by man starts with the Chinese. They are thought to have built **kites** more than 2000 years ago (Fig. 1.1). These kites are supposed to have been flown for religious reasons, but it is fairly likely that they flew them just as much for fun! They may also have used them for communication – as an early form of a telegraph system. The first man to fly was probably a condemned prisoner of the Chinese tied to a large kite. It might have been a test of his innocence if he landed safely! The historians have some evidence than man-carrying kites were used by Japanese warriors (Fig. 1.2) for spotting enemy armies and ships over the horizon.

There were two basic types – **man-carriers** and **man-lifters**. The man-carrier actually carried the man within the structure of the kite; the lifter used the kite as a 'sky hook' to lift the man suspended below the kite. Figure 1.2 shows a man-carrier.

The use of kites for scientific research started in Scotland in 1749 with Alexander Wilson. He measured the temperature at various altitudes by

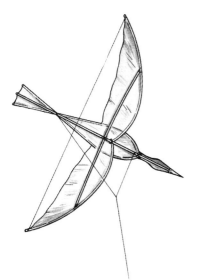

Fig. 1.1 A traditional Chinese bird kite

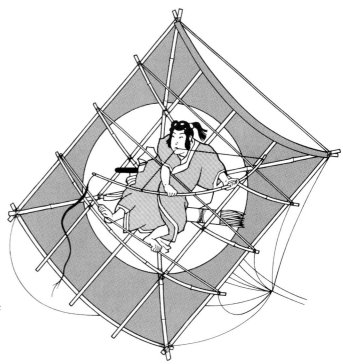

Fig. 1.2 A man-carrier

using thermometers strung on a *train* of kites. The very famous kite experiment of Benjamin Franklin was carried out in America in 1752. His kite was a simple flat cross-type tethered by a wet silk thread. The kite was flown into the thunder clouds and the electric charge was allowed to run down the string. The kite was also used to develop the first true **glider**, built by Sir George Cayley around 1800. He built a glider able to carry a man – its only flight was carried out in 1853. During the Victorian era many ingenious kite-like devices were built – to carry lines, to start the building of bridges, to rescue seamen in distress and to make measurements in the atmosphere. The kite was also commonly used for aerial photography as well as meteorological experiments. During the 1890s Lawrence Hargrave, an Australian, designed and built a series of model aeroplanes and kites. He designed a curved wing – the forerunner of the cambered **aerofoil**. His box kite is still flown today by enthusiasts who have built trains of man-lifters.

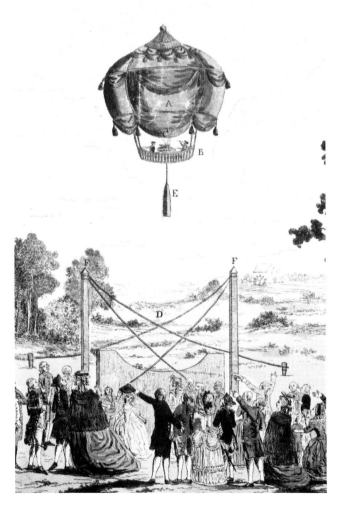

Fig. 1.3 The world's first aerial voyage (in a Montgolfier hot-air balloon) on 21 November 1783

■ Balloons and Airships

The first free flight by man was made in a **hot-air balloon**, constructed and tested by two brothers, Joseph Michel and Jacques Étienne Montgolfier (Fig. 1.3). It was in Paris in 1783. The brothers travelled 8 kilometres in the balloon. This was quickly followed by a 43 kilometre flight in a **gas-filled balloon** by another Frenchman, Jacques Charles, on 1 December 1783.

The balloon, however, depends upon the wind and must fly where the wind takes it. Modern hot-air balloonists can choose to fly at a particular altitude that may have wind at a particular speed or direction. The early balloonists did not have the information, the degree of control, or the accurate instruments necessary for this. They were at the complete mercy of the wind.

One way around the problem was to make the balloon cigar-shaped to reduce **drag** (see page 9), and to push it along with a draught of air from a rotating **propeller**.

Early attempts to drive the propeller using a steam engine (1852), and an electric motor (1884), were unsuccessful because they had insufficient power to overcome even the lightest breeze.

On 2 July 1900, a German, Count Ferdinand von Zeppelin, launched the first **Zeppelin airship** over Lake Constance. This was 130 metres long and cigar-shaped, with a blunted nose and stern. Below the craft were two 16 hp Daimler engines which gave the airship a speed of 30 km/h. This airship was cumbersome and slow but had some distinct advantages. It could stop in mid-air for essential repairs, and it did not tend to rotate like a balloon.

The gas used in gas-filled balloons and in Zeppelins was **hydrogen**. Hydrogen is the lightest, or least dense gas that there is. However, when mixed with air, hydrogen is very inflammable. This has resulted in some spectacular airship accidents involving considerable loss of life. Great Britain's *R101* crashed and caught fire in France on its way to India in 1929 and America's *Akron* crashed in the sea in 1933. When the great danger involved in the use of these airships was realised, many countries severely restricted their use. However, Germany, the birthplace of the airship, still placed more faith in the airship as a means of air transport than the aeroplane. This was because airships could then lift much heavier loads, remain airborne without motor power, and land in places where aeroplanes could not.

After 1918, Germany had been forbidden to build airships for military purposes as, during World War I, it had bombed London and British ports using Zeppelins. Nevertheless, in 1928 the civil airship, the *Graf Zeppelin*, was launched (Fig. 1.4). This flew until 1937, having flown nearly two million kilometres without accident. Also in 1937, Dr Hugo Eckener built the civil airship, the *Hindenburg*, which on 6 May 1937 suddenly burst into flames on landing in Lakehurst, New Jersey. Eckener had been worried by the potential danger of using hydrogen and had wanted to use another gas, **helium**, instead. Helium is slightly heavier than hydrogen but completely safe as it cannot burn. However, the main source of helium was America, who refused to sell Germany this gas. This was because the

Fig. 1.4 An airship – the *Graf Zeppelin*

Americans were worried that Germany might use it for re-armament purposes. For a long time after the *Hindenburg* disaster, airships were only used for limited work, such as coastguard duties in America. Their vulnerability to bad weather, lack of manoeuvrability, and the constant fear of accident made them much less attractive than aeroplanes.

■ Gliders and Planes

Neither the Chinese nor the Romans developed flight technology any further than the kite and, of course, you may not consider a kite as real flight, because it is always tethered to the ground in some way. The next stage in flight development was not until around AD 1500 when the inventor Leonardo da Vinci designed flying machines to be propelled by human muscle power (Fig. 1.5). He thought that man would be able to fly if he had wings the same shape as those of birds!

Fig. 1.5 Ornithopter design by Leonardo da Vinci, *c*.1485

Then came the **glider**, the forerunner of the aeroplane. In the nineteenth century model gliders were developed by Sir George Cayley. Cayley is now often referred to as the 'father of the modern aeroplane'. He carefully studied aerodynamics and in 1804 designed a glider which can be regarded as the first scientific aeroplane (Fig. 1.6). In 1849 Cayley built

Fig. 1.6 Sir George Cayley's model glider

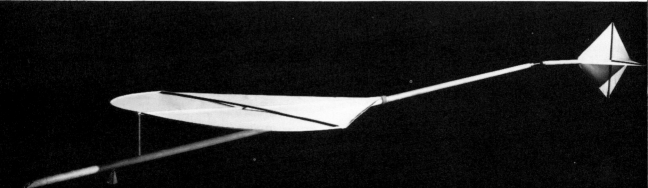

a glider that carried a boy for a short distance, and in 1850, when he was 80, he designed a glider that could carry a man. In Germany between 1890 and 1896, Otto Lilienthal made man-carrying **hang-gliders** which the pilot was able to control by shifting his weight from side to side.

The development of powered flight was not possible until a light, powerful engine was developed. This was the **petrol engine** which came into its own by the start of the twentieth century. Powered flight was achieved for the first time in 1903 when two brothers, Orville and Wilbur Wright, added a small petrol engine to a controllable glider. They were able to fly freely, without depending on the wind direction or lift from a low density gas. The petrol engine for the Wright brothers' aircraft had the highest power to weight ratio of any engine up until then. Since the beginning of this century, the aeroplane has been steadily developed in speed, size, range and reliability. This has sometimes been under the influence of war and at other times in the light of severe competition for profits. This development led finally to the first **jet plane** being built in 1937 in Germany by Pabst von Ohain. Fortunately, the German High Command ignored this development at the time. Also, the jet engine built a little later by Frank Whittle proved to be much better technically. It was then not very long before aeroplanes could fly faster than the speed of sound.

The first commercial jet aeroplane was the *Comet I* built by de Havilland. This went into service in 1952 with the British Overseas Airways Corporation, or BOAC. Many other airlines were forced to 'buy British' to remain competitive, and for a while Great Britain led the world in commercial jet flight. Then in 1954 accidents showed that there was a design fault and the *Comet I* was grounded whilst the fault was isolated and put right. This delay helped the Americans to take the lead in commercial jet flights.

☐ The Helicopter

In AD 1500, Leonardo da Vinci designed a **helicopter** but – as far as is known – he never actually built it. Cayley designed and built a model helicopter in 1800. This was able to fly up to about 30 metres into the air. Throughout the 1800s pioneers tried to produce enough power to turn the blades fast enough to achieve take-off and then sustain flight. The use of steam, electricity and even chemical explosions was considered, and in many cases tried, but with no success. In 1907, a Frenchman called Comu made a machine which lifted him and a passenger one and a half metres into the air for one minute. This used a 24 hp petrol motor. In 1909 Igor Alexis Sikorsky, a Russian-born, naturalised American, made his first helicopter, but it was too heavy to fly with the engines then available. In 1940 a German team lead by Focke used a 1000 hp engine in a helicopter which could climb to 7000 metres. Sikorsky followed this in 1941 with a machine that could fly horizontally, as well as climb vertically. However, helicopters have limited use as they are noisy and expensive to run. Air-sea rescue, short-distance inter-city travel, crop spraying and military use are the main areas in which helicopters are used.

■ The Aircraft Industry

The aircraft industry is of vital importance to many people for jobs, not only in the aircraft factories but also in the services that are necessary for the people who work there – the schools, hospitals, shops, and other service industries.

Aircraft and engines are a valuable export since they command a high price in relation to their weight, and represent a 'high value-added' product. Aircraft engineers depend upon previous experience in order to guide their decisions, and upon new discoveries made during research. They must work in a team since no one can possibly understand the whole field of aeronautics. The team consists of individuals who each contribute towards the design building and operation of that aircraft.

Britain still leads the world in several developments such as *Concorde*, vertical take-off aircraft and hovercraft. Other countries lead in different areas such as space travel.

2 Balloons and Kites

■ **Introduction**

This chapter and the next discuss developments that preceded the aeroplane and describe man's early attempts to become airborne.

■ **Balloons**

Balloons were man's first means of raising himself off the ground in free flight. Even now they cause a great deal of excitement whenever they are seen. Today balloons are used in sport (hot-air balloons) and for carrying out tests of the upper atmosphere for weather reports. They are also sometimes used as floating advertisements. During World War II **barrage balloons** (Fig. 2.1), which were tethered to the ground, carried cables to snare hostile aircraft over cities.

Fig. 2.1 A captive barrage balloon

Balloons in air work like submarines under the surface of the sea. A submarine can hover under the surface wherever its weight is exactly the same as the weight of its own volume of sea water. The submarine and its **ballast** have a total weight that is averaged out over the whole 'body' of the submarine. Decreasing the submarine's weight by pumping out ballast makes it rise, eventually to float on the surface. Increasing the submarine's weight by flooding in ballast, makes it sink deeper, eventually to rest upon the sea bed.

Gases and liquids have a collective name – **fluids**. Any **body** in a fluid can hover in the fluid if its weight is exactly the same as its own volume of the fluid. The word 'body' here doesn't mean a human body – necessarily! It means a thing, or a machine – anything we care to define. So it applies perfectly well to a balloon in air, even if the balloon is made up of the hot air or gas-filled lifting part, together with the load part attached underneath. All we have to do is remember to consider all of the parts of the body together so that we balance the **lift** against the **load**. When they are equal, the result will be a **hover**. If the lift is greater than the load, it will rise. And if the load is greater than the lift, either it will not take off or, if it has already taken off, it will start to come down. Consider a **hydrogen**-filled balloon in air with a volume of 1 m³ (Fig. 2.2). We know that:

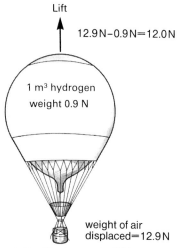

Fig. 2.2 Hydrogen's lifting power

1 m³ of air weighs	12.9 N
and 1 m³ of hydrogen weighs	0.9 N
so the lifting force of	
1 m³ of hydrogen is	12.0 N

This is well over 90% of the weight of the air displaced. However, hydrogen and air can form highly explosive mixtures. **Helium** is more expensive to buy and slightly heavier than hydrogen but it is completely safe. Let us see what lift we would get with helium. We know that:

1 m³ air weighs	12.9 N
and 1 m³ of helium weighs	1.8 N
so the lifting force of	
1 m³ of helium is	11.1 N

This is less than a 10% reduction on hydrogen's lifting force, still giving lift equal to over 80% of the weight of the air displaced.

The reduction in lift between helium and hydrogen is relatively unimportant. The problem is that helium is a rare gas, found mainly in natural gas deposits in America, and is expensive to separate from the other gases there. However, nowadays, considering the total costs of a lighter than air commercial flying machine, this extra cost is more than compensated for by the extra safety.

For fun flying by amateur balloonists, the development of inexpensive and fairly light propane burners has led to the enormous popularity of **hot-air balloons** (Fig. 2.3). We can carry out a similar calculation for a hot-air balloon to that for hydrogen or helium balloons. Consider a hot-air balloon with an average hot air temperature of 80°C in cold air at an average temperature of 0°C.

We know that:

1 m³ of cold air weighs	12.9 N
and 1 m³ of hot air weighs	9.9 N
so the lifting force of 1 m³ of hot air is	3.0 N

This is only about a quarter of the lift from either hydrogen or helium, but it is relatively safe and inexpensive.

The hot air tends to cool and diffuse, so the balloon loses its lifting power unless the burners are turned on from time to time to keep the balloon at the same height. Therefore **propane** has to be burnt from time to time during a flight. For a long flight quite a lot of propane may be needed. It is stored in steel cylinders in a compressed form. So the balloon has to lift the cylinders and burners as well as its normal load.

Modern amateur hot-air balloonists value the teamwork that is necessary to unpack, lay out and fill their balloons. On the ground, even in the lightest winds, the balloon must be controlled by a ground crew. Then, in the air, the crew experience the stillness as the balloon is carried along by the wind. They can hear every sound from the ground below and sometimes carry on conversations with the people they pass. Meanwhile, the ground crew follows on to help with the recovery and repacking of the balloon when it lands.

Fig. 2.3 A hot-air balloon

■ Airships

Between 1900 and 1938 hydrogen-filled **rigid-frame** airships were used to carry passengers over long distances (Fig. 2.4). The rigid frame was able to hold its fairly **streamlined** shape, and with the help of engine-driven propellers was able to move against a gentle wind. The hydrogen was held in bags inside the streamlined shape, and did not fill the complete shape. The airship was still difficult to control, particularly in high winds. Because hydrogen is such a lightweight gas and has very small molecules, it would gradually leak out of the gas bags by **diffusion**. It would have to be replaced from cylinders or hydrogen generators, or the airship would have

Fig. 2.4 The *Hindenburg* airship

to be lightened. Therefore every airship had to carry ballast to make sure that it could maintain height. Ballast was usually water which could be released from the airship without danger to anyone on the ground.

Airships had the ability to stop in mid-flight. This was essential to carry out repairs on the then fairly unreliable engines. Londoners tell the story of hearing someone hammering an engine overhead during a World War I bombing raid!

As has been mentioned earlier, airships were used both by Germany in World War I and, between then and World War II, by Great Britain, France, America and Germany for civil aviation. There were transatlantic flights and flights to the Far East. However, horrific accidents occurred with airships. The accidental destruction by fire of the *Hindenburg* was filmed by a cameraman. This film was shown to cinema audiences with the live and very emotional commentary made at the time. The *Hindenburg* was the last of the great hydrogen-filled airships.

After the *Hindenburg* disaster the public felt that airships were too dangerous, and commercial transatlantic airship flights were stopped. The full-scale use of helium did not start until after World War II, by which time aircraft were even faster, much safer and more comfortable. They did not suffer as much buffeting from winds as airships did, and the drag on an aircraft is much less than on an airship. Also by then engines had been developed that were reliable for many hundreds of hours of operation – so the ability to stop for essential repairs was no longer needed!

Recently, plans have been made to use airships as a paying concern once again. However, they are not likely to be used widely to carry fare-paying passengers or cargo because they are so affected by adverse winds and are unreliable for commercial purposes. They are more likely to be used for advertising or military purposes. Goodyear in America developed a series of helium-filled airships, their *Europa* entering the European market in 1972. It is used for advertising, for aerial survey work, and as a TV camera platform.

In Britain, a company called Airship Industries has been set up recently to expand the use of airships in such fields as fishery protection, long range radar, and promotion for commercial concerns. They built the *Skyship 500* in 1983 and the *Skyship 600* in 1984, and developments still continue.

■ Kites

Kites are really tethered aircraft with the air passing over them rather than them flying through the air. However, they differ from free-flying aircraft as the 'wing' surfaces do not have the special lifting shape of an aeroplane's wings. The air can swirl around the kite without it falling to the ground as an aeroplane would.

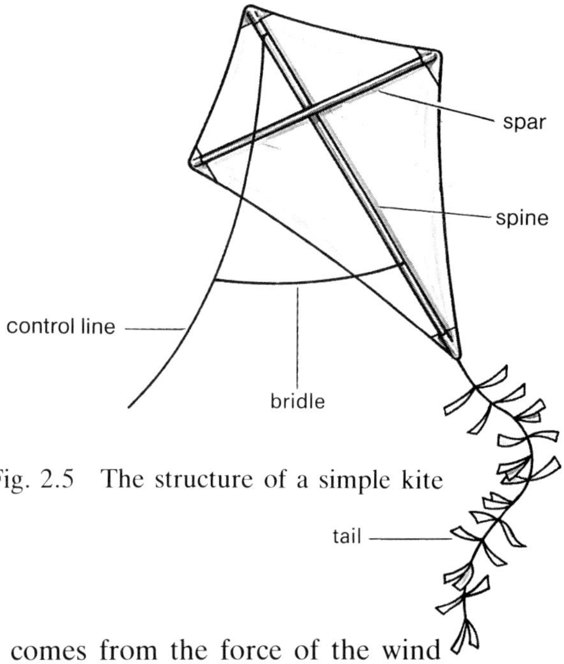

The wings of a kite are the large flat surfaces, often on either side of the central spine spar of the kite (Fig. 2.5). There is a limit to how large these can be made before they cease to be sufficiently rigid.

A kite flies as a result of the air flowing over and pushing against its lower surface, so giving the kite **lift**. The air does not flow smoothly over the kite's upper surface but swirls over it as can happen over an aircraft wing which has **stalled** and has lost much of its lift.

Fig. 2.5 The structure of a simple kite

The kite has to be adjusted so that it is balanced around each of the two lines (or **axes**) shown dotted in Fig. 2.6.

The **lift force** produced from a kite comes from the force of the wind upon its surface. There is also a **drag force** from the airflow round the kite. For a kite to stay in the air the combination of the lift and drag forces must be equal and opposite to the combination of the kite's weight and the tension in the control line (see Fig. 2.7). To lift a large weight the kite must have a large surface and must also have that surface at the correct angle to the wind to obtain the necessary lift.

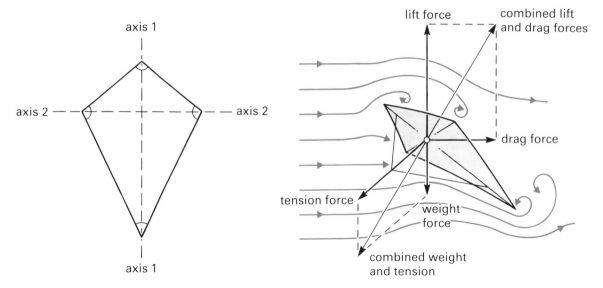

Fig. 2.6 Balancing a kite

Fig. 2.7 How a kite flies

A good kite design will have a lightweight structure, strong enough to withstand the forces exerted by the wind and capable of supporting the maximum lifting surface area.

A good example of the light strong structure is the **box kite**, provided for pilots shot down into the sea during World War II (Fig. 2.8). These kites were used to carry an aerial so that signals could be transmitted by radio.

Fig. 2.8 A 'downed' pilot uses a box kite to carry a radio aerial

Another structure which is light but strong and in which nearly all the kite is producing lift consists of polystyrene tiles arranged in triangles at each end of short, rigid support spars, as shown in Fig. 2.9. These short, rigid support spars will be less likely to bend than the long, slender spars of the kite shown in Fig. 2.5.

To keep the kite at the correct angle, the length of the cables used to **rig** it must be in the correct proportion. Also the kite should balance along its centre line – if one side is heavier than the other it will topple.

The kite has to be rigged correctly in order to fly with maximum efficiency. Sometimes drag would be required and the kite would be called a **puller**. But for most applications lift is more important, and a **floater** type of kite is required. These differences are illustrated in Fig. 2.10.

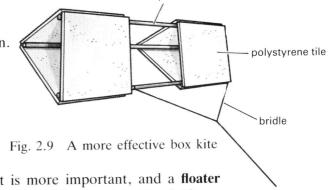

Fig. 2.9 A more effective box kite

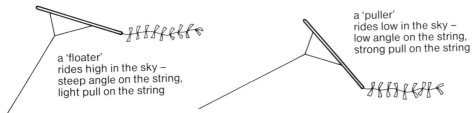

Fig. 2.10 A 'floater' and a 'puller'

If a kite has more than one control line, it will be more controllable – as in the **stunter kite** shown in Fig. 2.11. However, the single-string kite can be kept quite stable by adding a tail. This acts rather like an anchor, keeping the nose of the kite pointing into the wind (Fig. 2.12). The tail interacts with the air flowing past it, producing additional drag. This turns the kite so that it always faces in the right direction.

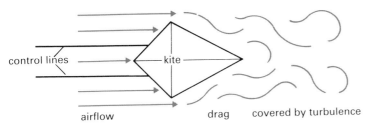

Fig. 2.11 A stunter kite has two control lines

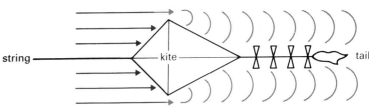

Fig. 2.12 The effect of a kite's tail

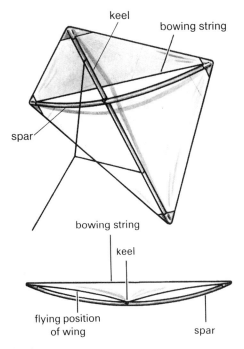

Fig. 2.13 A bowed kite

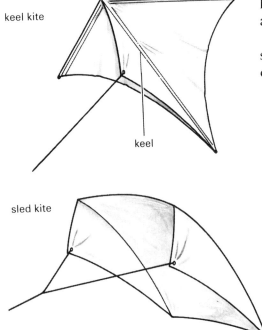

Fig. 2.14 Keel and sled kites

A bowed kite, with **dihedral**, is much more stable than a flat kite and does not need a tail. Dihedral is the angle formed by a pair of lift surfaces or wings. The action of dihedral is similar to that of a saucer floating on the surface of water. A bowed kite is shown in Fig. 2.13. We shall learn later that aircraft wings are often arranged to have dihedral, in order to provide stability. **Keel** and **sled** kites also avoid the need for a stabilising tail. These kites are not rigid but flex with the wind. They are shown in Fig. 2.14.

The control line also experiences drag from the wind. This helps explain why the line tends to sag so much (Fig. 2.15). Biplanes with wire rigging between their wings suffered similar drag and this is one reason why single-winged monoplanes became more popular.

A kite has to lift itself and its string. The lighter the kite, the more string it can lift – so that it can fly higher. However, the lighter the material of the kite, the weaker it is. Therefore kite design is a compromise between strength and weight.

Kites are fun to make and fly but there are some precautions you ought to take if you go out with a kite. Never fly a kite:

(a) near a road or railway.
(b) within 8 kilometres of an airport.
(c) in a thunderstorm.
(d) near overhead power lines.

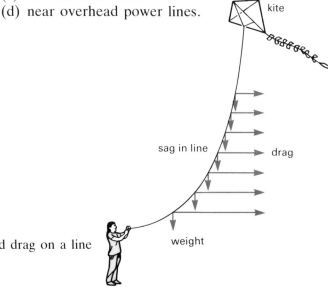

Fig. 2.15 Wind drag on a line

If you can find safe areas, it is interesting to compare a kite's flight in the following places:
   (a) on the seashore with a breeze blowing towards the land;
   (b) on steep hillsides, taking great care (Fig. 2.16);
   (c) near trees or near buildings (Figs. 2.17 and 2.18);
   (d) on a flat exposed area of land.

In areas where there are few obstructions to the smooth flow of air a stable kite can be tethered, for instance to a gate post, and left to fly by itself. In other places the kite is tossed to and fro due to eddies and swirls of air called **turbulence**. These are caused where obstructions force the air over and around them. Sometimes this turbulence is too rough to fly a kite. Aeroplane pilots especially have to try to avoid flying too low over hills and high buildings because the turbulence can throw the aircraft off course. If such turbulence occurs when the aircraft is too close to the ground, as when taking off or landing, the pilot may have great difficulty. This is one reason why flights are cancelled if there are very high wind speeds.

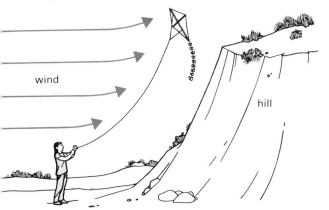

Fig. 2.16   Flying a kite in the steady updraught near a hill

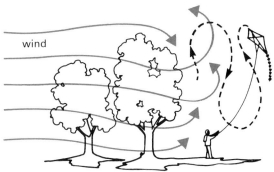

Fig. 2.17   Turbulence behind trees makes kites unstable

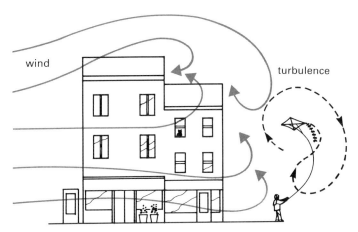

Fig. 2.18   Turbulence behind a building makes kites unstable

3 Gliders and the Wright Brothers

Introduction

The first real glider was designed in 1809 by Sir George Cayley (Fig. 3.1). From his aerodynamic studies he decided that the flapping wing machines designed by earlier pioneers should be replaced by a **fixed-wing** aircraft. He built stable gliders with tail surfaces to control the aircraft. The design of his gliders was the forerunner of the shape of aeroplanes to come. The position and shape of the rudder, tail plane and wing are the same as those of today.

During the later part of the nineteenth century, curved wing surfaces which produced aerodynamic lift (gliders) were patented. However, a German engineer, Otto Lilienthal, was the first man to fly any distance in a glider. His gliders are properly called hang-gliders – controlled by the aviator shifting his weight from side to side (Fig. 3.2). This altered the craft's centre of gravity and so produced a movement in the same direction as the weight shift. This principle is still used today in modern hang-gliders. Lilienthal and his follower, Percy Pilcher, were both killed in gliding accidents. Lilienthal stalled at a low altitude and was unable to recover. In the five years he had been flying he made over 2000 flights, and had achieved flying distances of over 300 metres.

Fig. 3.1 Sir George Cayley's man-carrying glider

Fig. 3.2 Otto Lilienthal's hang-glider

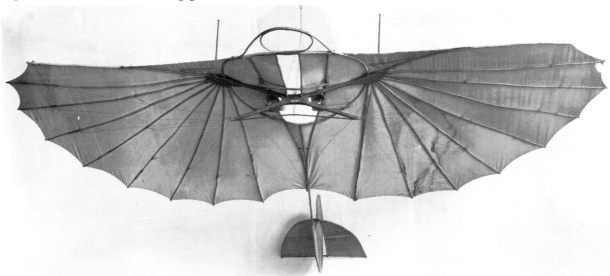

The Wright Brothers

The men who eventually mastered the techniques of glider flight were the Wright brothers. Orville and Wilbur Wright were the sons of an American bishop. When aged 10, Wilbur was given a toy, powered by a rubber band, which could fly to the ceiling. Fascinated by this toy, Wilbur made other models like it. He also made bigger models but found that these would not fly. From this he learnt that as a flying machine is made bigger, its weight increases more than its strength. The two brothers started to earn a living by selling bicycles, and during the slack winter months spent their time and money building gliders. In 1900 they built a workshop at Kitty Hawk, an area of sand dunes in Northern Carolina where the wind blew in a steady and regular way. Here they built and tested their first glider, and two years later made the first **powered flight**. This was on 17 December 1903 (Fig. 3.3) and was an historic occasion. Orville Wright wrote:

> 'With all the knowledge and skill acquired in thousands of flights in the last ten years, I would hardly think today of making my first flight on a strange machine in a 21 m.p.h. wind, even if I knew the machine had already been flown and was safe . . . Yet faith in our calculations, and the design of our first machine, based on our table of air pressures, obtained by months of careful laboratory work, and confidence in our system of control, had convinced us that the machine was capable of lifting itself into the air and that, with a little practice, it could be safely flown . . . Nevertheless, it was the first flight in the history of the world in which a machine carrying a man had raised itself by its own power in full flight, had sailed forward without reduction of speed, and had finally landed at a point as high as that from which it had started.'

Fig. 3.3 The Wrights' first plane – *Flyer I*

The two brothers had done much more than carry out the first powered flights – they had assembled and checked the information then available for the study of aeronautics. When information was unclear or not available, they had compiled a body of accurate knowledge from scientific observation.

There were three problems that faced them: firstly, how to get enough lift on the wings; secondly, how to control the aircraft; and thirdly, how to obtain a lightweight petrol engine with enough power to propel the aircraft when carrying both the pilot and the engine.

The first problem was solved by building a **wind tunnel** to test the lift from small curved wings. Using delicate balances, they were able to measure the forces on these wings and so find the curved shape that gave the biggest lift.

To solve the second problem, the Wrights realised, from a careful study of Lilienthal's work, that his crash had been caused because he had failed to provide for some way in which the wings of his gliders could be **warped** or twisted – as birds twist and turn the edges of their wings to maintain balance. The Wrights warped their wings by having the pilot pull wires that stretched through the wings.

They solved the third problem by designing and building their own **petrol engine**. This had a weight of only 70 kilograms but gave 12 hp. They used it to turn two propellers placed behind the wings of the aircraft to pull it forward just like ships' propellers.

The Wrights also learnt to bank by increasing the wings' angle to the forward direction, the 'angle of attack'. As we know today, this produces more lift. They later added a rudder and a forward elevator to give control in the pitch and yaw planes. These are described in detail later in this text. Their aeroplane was effective and worked very well and they continued with its development (Fig. 3.4).

Fig. 3.4 The Wrights' later biplane in flight over France in 1908

World Wars I and II

The story of the glider was overshadowed by the development of the powered aeroplane, both by the Wright brothers and their peacetime successors, as well as by World War I. However, after that war, the Germans were unable to obtain powered aircraft due to treaty obligations. This led to further work on gliders by the Germans. They developed gliders which were very light and which were launched by taking them towards the airflow coming up a sharp incline. This was called a bungee tow (Fig. 3.5).

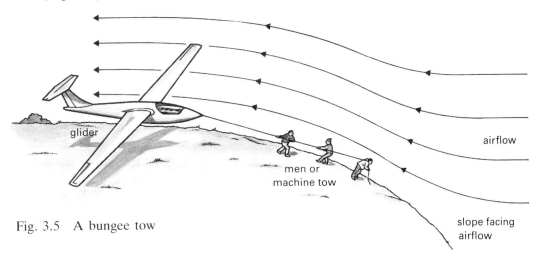

Fig. 3.5   A bungee tow

Gliders such as in Fig. 3.6 gave the German airforce, the Luftwaffe, air experience at a time when powered aircraft were not available to them. Pilots gained experience of military flying for when Germany rearmed.

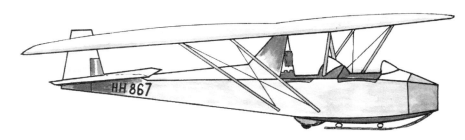

Fig. 3.6   A World War I glider

Gliders were used during World War II as a low-cost means of transport for men and materials. They had the advantage over powered aircraft of being able to land in small fields due to their low landing speeds. If damaged, they could be repaired at a low cost or, if recovery was too difficult, they could be left there as 'disposable' aeroplanes. In Germany a large glider of about 60 metres wing span was designed and built to carry 130 fully-armed troops. This was the ME 321 *Gigant*. These were towed up by three powered aircraft (ME 110s) but were slow, massive (40 tonnes) and difficult to control. Eventually these gliders were converted to powered aircraft by the addition of six engines.

The most famous use of gliders in World War II was in the Allied invasion of Europe on 'D' day, and in the battle of Arnhem in Holland. Here they were used to carry Allied forces behind enemy lines in order to cut off the retreat of a large portion of the German Army. Heavy bombers towed the gliders to the area of the 'drop'. After releasing them, the bombers would return to base, leaving the glider pilots to fly the fully-laden gliders to what could only be described as a hazardous landing!

■ Modern Gliding

The glider uses the upward flow of air from hills and the rising flow of air in **thermals** (Fig. 3.7). These are areas of rising air produced by the heating effect of the sun on the ground. The glider – having no engines – must always be gliding downwards through the air, but if the air is moving up faster than the glider moves down, the glider will climb.

The glider gains its altitude initially with a tow from either a fixed winch or a powered aircraft. When the glider reaches a particular altitude, the pilot releases the tow cable and goes into a glide. If there is no upward flow of air, the aircraft will glide towards the ground. The angle of glide depends upon the lift from the wings and the drag on the aircraft. This is called the **lift/drag ratio**.

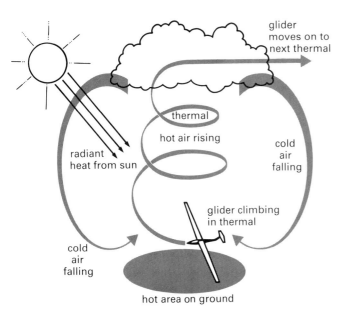

Fig. 3.7 Gliding and thermals

As a complete contrast to the gliders we have discussed so far (all rigid structures) there is the hang-glider. During the 1950s a NASA engineer, Francis Rogallo, developed a flexible-wing kite which, totally unsupported, was inflated by the wind and kept in shape by the shroud lines (Fig. 3.8). This principle is now used in parachutes where skydivers wish to alter their direction as well as reduce altitude slowly. But the most popular development of Rogallo's is the **hang-glider**.

This is basically a flexible kite (Fig. 3.9) with wings that flex according to the flow of wind. It is a **delta wing** shape, supported by three longitudinal aluminium tubes and a **strut** to keep the delta shape. The flier hangs below the wing and controls the glider by moving his or her body from side to side – just as with Lilienthal's gliders.

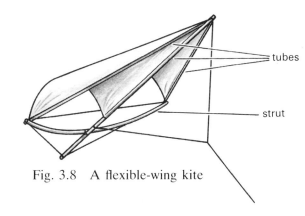

Fig. 3.8   A flexible-wing kite

Fig. 3.9   A modern hang-glider

All aircraft, powered and unpowered, are designed to glide (Fig. 3.10). For powered aircraft, this is so that if their engines cut out they will not stall and will still have forward motion. In a glider, where duration of flight rather than speed is the main consideration, the gliding angle is made as small as possible. However, with a powered aircraft other requirements usually result in the gliding angle being considerably bigger than that for a glider.

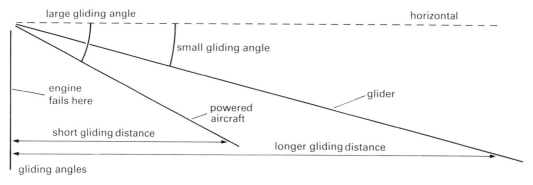

Fig. 3.10 Gliding angles

■ The Gliding Angle

The weight force of a glider can be imagined as being split into two parts, called components. These are shown in Fig. 3.11 as Wx and Wy. The strength of the two components, Wx and Wy, can be found by using the parallelogram of forces. Applying this technique to Fig. 3.10, the **weight force** is drawn to scale as CA, and the **drag force** as CE. Wx is then drawn the same length as the drag force, but in the opposite direction, as CB. Wy is then obtained as CD when the rectangle ABCD has been completed. At any given steady speed and angle of glide the weight force component Wy is balanced by the lift, and the weight force component

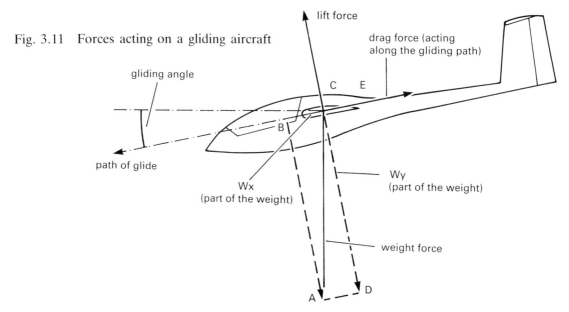

Fig. 3.11 Forces acting on a gliding aircraft

Wx is balanced by the drag force. If the drag force is high, the gliding angle will need to be large and the glider will lose height quickly (Fig. 3.12). Drag is kept as small as possible to keep the gliding angle low.

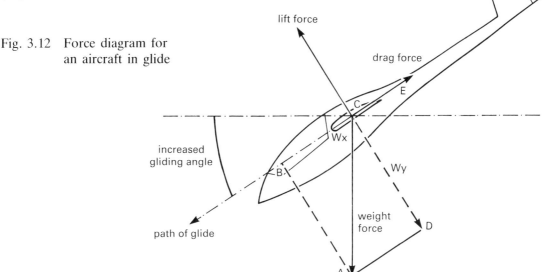

Fig. 3.12 Force diagram for an aircraft in glide

This will give the pilot the maximum time aloft with the best chance to take the most advantage of **thermal currents**. When the pilot finds a thermal current, he circles within it to gain as much altitude as possible. When the pilot has gained sufficient altitude, he can navigate across country, gliding at as low an angle as possible, and all of the time looking for the next thermal or updraught. It is almost like a game of snakes and ladders! It is very important for a glider to move easily through the air. Drag should be avoided as much as possible.

A competition glider has all of its surfaces shiny and smooth. Such aircraft have to be carefully handled to avoid getting grit on their surfaces. Even a small amount of dust, say 0.01 mm thick, can make the difference between soaring and sinking. The skin is highly polished, and all screw heads are countersunk. The frontal area of a glider accounts for a large proportion of the glider's air resistance, or drag, and the position of the pilot in the cockpit has considerable influence on this area. By adopting a semi-reclining position the pilot can reduce the frontal area and thus the drag (Fig. 3.13). Designs are often tested in wind tunnels.

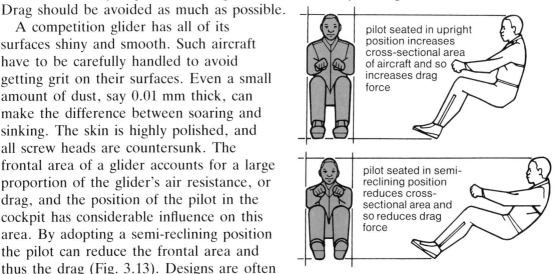

Fig. 3.13 Reducing the cross-sectional area

Wings produce drag due to **vortices** or whirlpools in the air at their tips. This is called **induced drag** and is reduced by making the wing span very large, and the width of wing very small. This is discussed in greater detail later on. The average distance between the leading and trailing edges is known as the **mean chord**. The ratio of the wing span to the chord is called the **aspect ratio** (Fig. 3.14). A high aspect ratio wing is necessary in a glider where we want low induced drag with high lift, i.e. a high lift to drag ratio.

It is very useful to engineers to use ratios like aspect ratio and lift to drag ratio whenever possible. This is because different designs can then be compared independent of their actual size.

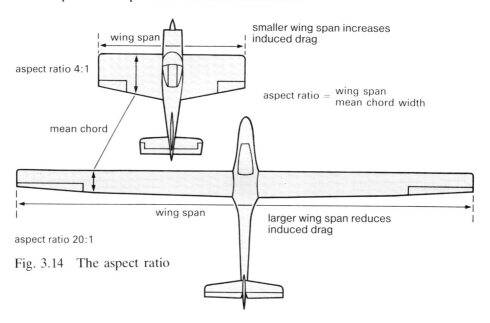

Fig. 3.14 The aspect ratio

■ Gliding as a sport

Today gliding is a popular sport. Gliders are flown at high altitude hundreds of kilometres across country. Many people join a gliding club or participate in Air Scout or ATC activities as their first introduction to flying. Few people can afford to own a glider by themselves and very often several people join together to form a syndicate to pay for and operate a glider between them. As with balloons, the members of the club do much more than fly. Gliding is a team sport where members work together in order to launch the glider for flight, recover it in due course, and maintain it in good flying order.

Hang-gliding with its lower capital costs has also become very popular. There are probably far more people now involved in hang-gliding as a sport than there are involved with rigid-structure 'conventional' gliders.

4 Aircraft Structures

■ Introduction

This chapter deals with the way in which aircraft are built and the problems that face aircraft designers.

Figure 4.1 shows cross-sections of the bones of a cow and bird. The cow bone is:

(a) thick walled;
(b) heavy;
(c) very strong to support its weight.

The bird bone is:
(a) thin walled;
(b) light;
(c) still strong and stiff.

Fig. 4.1   Bone structure

(a) Cow

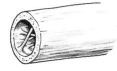
(b) Bird

A bird has a small weight but still requires strong, stiff bones. The strength and stiffness of the bony structure is needed to keep the bird in its proper shape; but the weight of the bony structure is not needed for any purpose. The weight therefore is said to be **redundant** and the bone structure should be as strong and stiff and lightweight as possible.

It is important to note that strength alone is not sufficient. Many structures are strong enough not to break, but bend or deflect too much because they are not stiff enough.

Aircraft structures need to be lightweight, strong and stiff. They need to be lightweight to allow the aircraft to fly. In the commercial aircraft business the lighter the airframe, the more cargo or passengers can be carried. The structure must be strong and stiff to carry the cargo or passengers in safety without distortion of the airframe.

■ Safety Factors

Aircraft designers have to consider safety factors. A safety factor of 50% means that if the aircraft is meant to carry a load of 10 000 N, it should withstand a load of 10 000 N plus 50% of 10 000 N, i.e. plus 5000 N. Therefore:

maximum load = 10 000 N + 5000 N = 15 000 N

When a load is accelerated, it can exert a greater force than its static weight. Therefore loads have to be stated for dynamic conditions, not static ones. Allowance must be made for the effective weight of a load when the aircraft is turning or diving rather than just for level flight. When it is accelerated, a load effectively increases its weight. On or near the surface of the earth a freely falling body will accelerate initially at 10 m/s^2, and a stationary body will weigh 10 N/kg. This normal gravitational effect is often called g. A load with a 1000 kg mass accelerated at a rate equal to g would have a weight of 1000 kg × 10 N/kg or 10 000 N. If it was accelerated at 2 g (i.e. 20 m/s^2), it would effectively have a weight of 1000 kg × 2 × 10 N/kg or 20 000 N. As an aeroplane's safety factor is increased its structure becomes heavier. The aircraft will then not be so attractive to buyers because it will not be able to carry as much cargo. Savings in weight are made by improvements in design rather than by cutting down on safety.

■ Stiff and Light Structures

Until aeroplanes were built there was no need to build light, stiff structures. Up till that time, machines which provided motion, such as steam locomotives, cars and ships, were massive and heavily constructed with iron and steel. Both the building materials (iron and steel) and the fuels (coal and oil) were cheap. Aluminium, which has a high strength to weight ratio, was relatively far more expensive than it is today, and was not generally available. Early aircraft were built using spruce – a light, stiff timber – and bamboo and wire. The **fuselage** (the body of an aeroplane) and wing surfaces were covered by canvas.

All forces can be considered as pushes or pulls, or as a mixture of pushes and pulls. A pull results in expansion or tension, and a push results in compression or squeezing. Early aeroplane builders used steel wires to provide tension to balance expansion and timber struts to resist bending and compression. Thus early aircraft makers used the skills of other construction experts.

The sailmakers' expertise was applied for the wings, that of carpenters for the frames and that of riggers to adjust the tension on the wires pulling the frames together. The aircraft were light and fairly rigid and were able to withstand the stresses put upon them. The wooden frame nonetheless had the additional advantage that it had some flexibility – it would 'give' rather than break suddenly.

The early wooden aircraft had four long wooden struts called **longerons** which ran the length of the fuselage (Fig. 4.2). Other shorter struts ran between the longerons to keep them apart and form a box shape.

Fig. 4.2 The fuselage structure of early aircraft

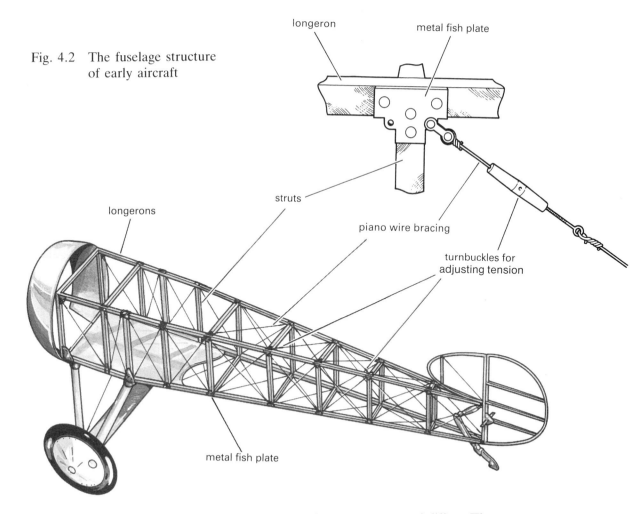

Stringers made from piano wire under tension gave extra rigidity. The covering fabric was fitted closely to this frame and was given a special treatment to shrink it to size; waterproof it and make it rigid (see page 34, Fig. 4.3). This treatment was called **doping**.

Later, wood was replaced by welded metal and metal wires by diagonal bracing tubes. As this method of construction was stronger, the resulting structure (main structure) was more open and stringer supports (secondary structure) were still needed to keep the fabric in the required shape. This method was in general use until World War II. Both for protection and pressurisation, most aircraft now have metal skins. However, fabric-covered aircraft are still flying regularly, the *Auster* and *Tiger Moth* being the most popular (see page 34, Fig. 4.4).

■ **Metal Structures and Biplanes**

The first all metal aeroplane was the *Monocoque Duperdessin*. Here an unbroken box structure was used, with curved sides. This gave strength, rigidity and lightness. However, any opening in this sort of structure

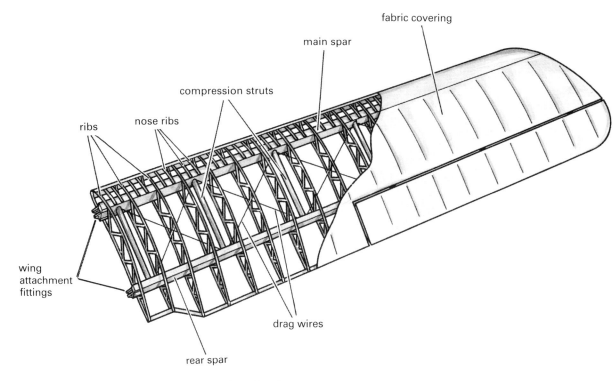

Fig. 4.3 An early aircraft wing

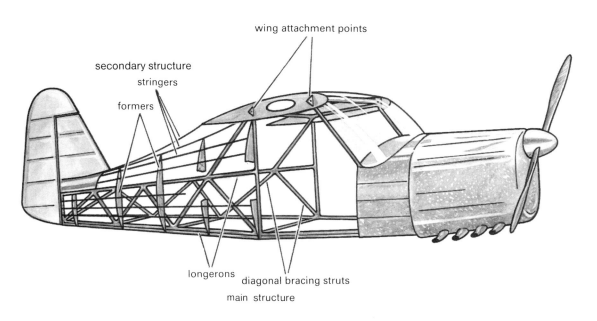

Fig. 4.4 The frame construction of an *Auster* aeroplane

weakens it considerably so a combined structure was evolved. Some stress was taken by the surface, which was stiffened and kept in shape by longerons and frames with extra stiffening near openings and wing roots.

At this time the biplane wing structure became one of the most common wing arrangements, as this structure is itself a braced girder. It is robust, even though the materials can be very simple. It also gives the aeroplane a large wing area. A monoplane with the same wing area was thought to be rather too flexible – the aeroplane would not have been able to make a sharp turn very easily. Manoeuvrable fighter aircraft with three wings, such as the *Fokker* triplane of Manfred von Richthofen, followed in World War I. These aircraft were very manoeuvrable, able to take off and land slowly, and able to climb quickly. During World War I, the German firm of Junkers developed aircraft built from **duralumin**, an aluminium alloy with much improved strength and only slightly greater weight than the other alloys. Other manufacturers were unwilling to change and started using this metal only as cowlings for engines.

■ Monoplanes

In 1935 Junkers produced an important design for an all metal transport monoplane. Some of the load was taken by the skin, which was corrugated – with the frame beneath taking the rest of the load. The wing was a cantilevered monoplane with the materials in it providing the resistance to bending. A **cantilever** is a structure where the weight of the lever is held firmly at one end (Fig. 4.5).

The strength of the material holds the structure rigid. The central part of the wing structure was a beam which connected the left and right wings through the fuselage.

Monoplanes became more and more popular, but up to World War II there were still biplanes in military use – *Swordfishes* at the Battle of Taranto, *Gloster Gladiators* on Malta. However, the smooth, low drag

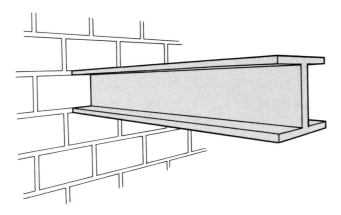

Fig. 4.5   An example of a cantilever

outline of the monoplane eventually spelt the end of biplane fighters. Smooth metal-covered wings became common but fabric-covered fuselages were still used, for example the *Hawker Hurricane*.

Some novel ideas were used in World War II – a plywood structure in the *Mosquito* and a geodetic structure (Fig. 4.6) in the *Wellington*. A **geodetic structure** has each of its parts in tension or compression and following the shortest line on a mathematically derived surface. This gives a very high strength to weight ratio. These were successful designs at the time but were not developed further.

Fig. 4.6 The geodetic airframe structure

☐ Pressurised Aircraft

Aeroplanes were designed to fly higher and higher – for smoother air conditions and optimum operating efficiency. Above about 3000 metres there is insufficient oxygen in the air for comfortable breathing. Test pilots could wear oxygen masks, but paying passengers could hardly be asked to do this too! In 1938 the Boeing company **pressurised** the whole passenger and crew compartment of their high-flying Boeing 307 *Stratoliner* – the first cabin-pressurised airliner. They increased the air pressure inside the aircraft to maintain an adequate supply of oxygen as the plane climbed to higher altitudes. It was not maintained at the equivalent to ground level, but it was increased to a comfortable level for everyone to be able to breathe without the use of oxygen masks.

The aeroplanes of World War II, flying high to avoid gunfire, were not generally pressurised. In other words, the air inside the aircraft was at the same pressure as outside. This meant that the aircrew had to be supplied with oxygen through face-masks when the aircraft was flying at high altitudes.

After the war the era of mass transport of large numbers of people began. Aircraft were also beginning to be powered by the jet engine which is more efficient at high altitudes. The problem now was that the aircraft structure had to contain air inside at a higher pressure than the outside, as well as resist the forces caused by the weight of the aircraft and its cargo. In addition to the pressure shell (or hull), the new pressurised aircraft had a new design featuring a strong metal outer skin fixed to ribs and stiffeners. This was known as a **stressed** skin. Figure 4.7 shows part of a wing incorporating this design.

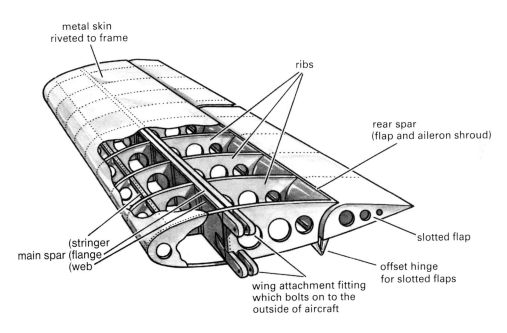

Fig. 4.7 The structure of a stressed skin wing

■ Safety and Fatigue

In Britain, de Havilland designed and flew the *Comet 1*, the first commercial jet aircraft which carried passengers. Unfortunately, after a period of successful operations, there were three unexplained crashes. All the remaining *Comets* were grounded. Systematic detective work identified changing loads on the structure as the aircraft took off, climbed and landed, due to the outside air pressure always changing. There were also changing loads on the structure due to buffeting from air currents. These effects combined to cause metal fatigue resulting in cracks in the stressed skin originating at the sharp corners of the window frames and spreading over the structure so that total failure occurred and the aircraft, as it started to break up, plunged to destruction (see Fig. 4.8).

The safety factors that had always worked until then were for the static load conditions experienced by existing aircraft. The **cyclic load** conditions encountered by the *Comet* aircraft structure were much more severe and the number of reversals of load built up more quickly due to the increased speed of the aircraft. Thus aircraft structures would fail at much lower loads than expected from slower, unpressurised aircraft.

Later designs avoided sharp corners (stress raisers) by making window and door openings oval in shape. The later marks of the *Comet*, such as the *Comet IV*, incorporated these improvements and flew successfully for many years.

The life of most aircraft and aircraft engines is limited by fatigue. Designers have to satisfy safety factors and make sure that the **fatigue life** of aircraft provides an economic number of flights.

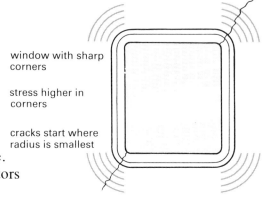

Fig. 4.8 A *Comet I* window

Regular checks are made of aircraft for the small cracks that precede fatigue failure. Aircraft are taken out of service if any cracks are found, and the still safe but failing parts are removed and replaced. Even proven designs can develop unforeseen flaws: for instance at one time some *Boeing 747s* were found with small cracks where none had been expected. That resulted in all *Boeing 747s* being taken out of service for a time, examined and repaired. It also led to modifications in design and in the routine inspection procedures.

At **supersonic** speeds, i.e. speeds faster than the speed of sound, fatigue cracks can start at any of the many rivets that hold the skin sheets to the ribs of an aircraft. In a subsonic aircraft the whole skin is held on in this way. However, *Concorde* has a fuselage made from machined sections with ribs included in part of the sections (Fig. 4.9). This reduces the number of rivets and other fixings. This method of construction is safer but is very wasteful as so much of the material is cut away during manufacture.

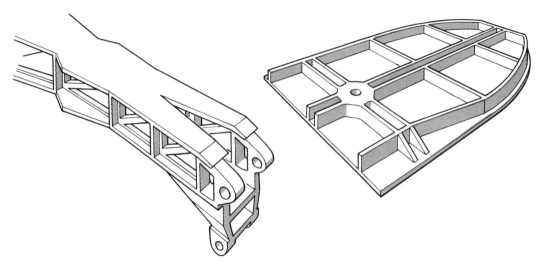

Fig. 4.9 Machined sections

☐ Modern Alloys

Most of today's aircraft are made from metal or **alloys** – mixtures of metals that combine their desirable qualities and minimise their non-desirable features. High strength and low weight are desired. Duralumin is an aluminium alloy now much preferred to aluminium itself. In addition to various steels, titanium is sometimes used because it keeps its strength at the higher temperatures so often created at supersonic speeds.

When an aircraft travels supersonically, air molecules hit the aircraft at such a speed that they raise the skin temperature. There are two ways of preventing the aircraft from suffering from this heating. The first method is by making the skin material resistant to high temperature – by using titanium or its alloys, or, as on the space shuttle, by using foam ceramic tiles. The second method is to cool the material from the inside of the aircraft so that it can be used at high speeds.

When designing an aircraft structure, the engineer considers the type of aircraft required and the type of operation which is required of the aircraft. The engineer applies limits and standards based on past experience and complies with regulations and proving tests set out by the aircraft control authorities – in Britain the Civil Aviation Authority (CAA).

Every new aircraft structure has to be tested. Nowadays aircraft are tested under extremes of pressure, temperature and load far greater than they will have to withstand in routine operation. Vibrations are applied to simulate dynamic flying conditions. These loads are applied and removed many millions of times until the structure fails. If the aircraft is found to fail these tests unexpectedly early, the structure has to be re-designed. All of these tests must be passed before the aircraft is given a licence to fly commercially.

5 The Principles of Powered Flight

■ **Introduction**

A modern aircraft may have a mass of over 350 tonnes or a weight of over 3.5 million Newtons! In order for it to fly, its forward movement must produce sufficient lift to overcome this weight. This chapter deals with how this is achieved.

If you watch a wide stream entering a narrow strait you can see the water increasing its speed. This effect can also be demonstrated by means of a **Venturi gauge** as in Fig. 5.1. If air from a blower is sent down the tube as shown, the water level in the central tube (B) rises more than in the tubes on either side (A and C). The water level in the vertical tubes indicates the pressure of the air as it flows through the horizontal tube.

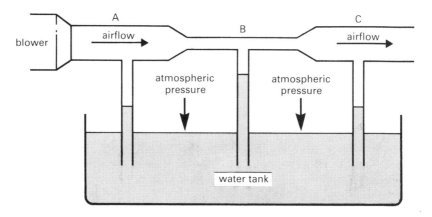

Fig. 5.1 Venturi gauge experiment

Atmospheric pressure acts on the free surface of the water in the tank so the raising of the water levels in the tube indicates a drop in air pressure, this reduction being caused by the flow of air. Thus the greater rise of water in the middle tube must be because of the even greater decrease in the pressure of the air flowing through the narrow part of the tube at B, than in the air flowing through the wider parts of the tube at A and C. The air flowing through the narrow part of the tube must therefore be travelling faster than the air in the wider parts. This fact is embodied in a statement known as **Bernoulli's principle** which is concerned with the pressure exerted by moving fluids – gases or liquids. This states that when the speed of a fluid increases, the pressure it exerts decreases; and when the speed decreases, the pressure it exerts increases. This principle is fundamental to aircraft wing lift as the shape of an aerofoil is designed to increase air speed over its upper surface.

■ Lift produced by an Aerofoil

We can see from Fig. 5.2 that a flat plate in a streamlined flow of smoke will cause a lot of disturbance and a curved **aerofoil** section very little disturbance. The smooth undisturbed flow of air round the aerofoil illustrated by the smoke streams is called **laminar** flow. The disturbed, swirling flow of air round the flat plate is called **turbulent** flow.

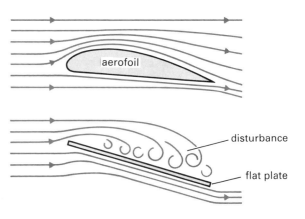

Fig. 5.2   Airflow round an aerofoil and flat plate indicated by smoke streams

The largely laminar flow round the aerofoil is necessary to produce lift efficiently. If anything happens to break up this flow into really turbulent flow, lift is lost and the aircraft can stall.

The crowding of the smoke lines above the aerofoil is similar to the crowding of the air that takes place in the narrow part of the Venturi gauge of Fig. 5.1. This shows that the air speed above the aerofoil is greater than the air speed below the aerofoil. Thus the pressure above the aerofoil, P_A, is less than the pressure below the aerofoil, P_B. Pressure equals force per unit area or force equals pressure multiplied by area, so this gives a net upward pressure of $(P_B - P_A)$, as shown in Fig. 5.3.

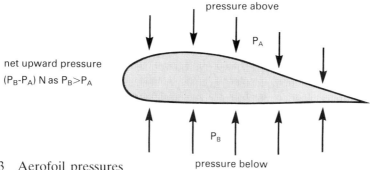

Fig. 5.3   Aerofoil pressures

□ Other Examples of Bernoulli's Principle Relevant to Aeronautics

The petrol engine carburettor works on the Bernoulli principle. The airflow is constrained above the injector nozzle from the petrol tank

41

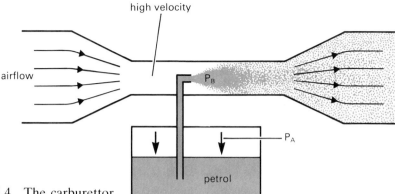

Fig. 5.4 The carburettor

(Fig. 5.4). Therefore the pressure, P_B, in this part of the carburettor is less than the pressure, P_A, above the petrol. This means that petrol is pushed into the airflow. It emerges as a fine spray, giving the right fuel to air mixture for combustion in the piston chamber.

The action of hovercraft also depends upon the Bernoulli principle (Fig. 5.5). The hovercraft's engines force air out under the craft at a high velocity, thus reducing the pressure, P_B, below the craft in comparison to the atmospheric pressure, P_A, above and around the craft. So P_A is greater than P_B and there is therefore a net downward thrust! This holds the hovercraft down onto the ground or the water over which it is operating. This limits the escape of air from underneath the craft and effectively forms an airbed over which the hovercraft can glide.

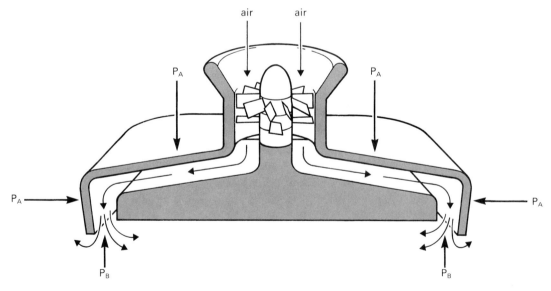

Fig. 5.5 A cross-section of a hovercraft showing air drawn in by a rotating propeller, and compressed under the hull to form a cushion of air for the hull to rest on. According to Bernoulli's principles, the pressure round the base of the hull is lowered, thus preventing the hovercraft from taking off.

Sir Christopher Cockerell was the first to put this to practical use. Figure 5.6 shows the first two hovercraft to be built, the *SRN1* in 1958, and the *SRN2* which went into passenger service in 1963. The *SRN1* was 10 metres long, operated at a height of 30 centimetres, and had a speed of 43 kilometres per hour. The *SRN2*, five years later, had a mass of 27 tonnes and a maximum speed of 140 km/h.

Fig. 5.6   The *SRN1* and *SRN2*

■ Lift Produced by an Aircraft Wing

The airflow around an aircraft wing has already been discussed and illustrated in Fig. 5.2. This shows a simplification of what actually occurs. This is because, however carefully an aerofoil is designed, small whirlpools form in the air over the wing surface and in the wake of the wing tips. These are called **vortices**. However, this simplified description will suffice here for our purpose. The lift is produced by the whole wing, but, as with any force acting over a rigid body, this may be represented as a single force acting on just one special point in the body. This point is called the **centre of pressure**.

If we consider the aircraft in level flight, the lift of the wings acting from the centre of pressure must balance the weight of the aircraft. For a level flying aircraft to be able to rise, the lift must be greater than the weight, and for the aircraft to descend, the lift must be less than the weight. The weight of the aircraft acts through the centre of gravity of the plane. This is the point about which the aircraft would balance. It is also the point about which it would rotate.

If the centre of pressure is forward of the centre of gravity, the aircraft will tend to tilt its nose upwards when attempting level flight. This is due to the rotating effect of the two forces, the weight and the lift. Two such forces make a **couple** and can cause a body to tilt or rotate.

If the centre of pressure is forward of the centre of gravity and the aircraft suddenly reduced its speed due to an engine failure, its nose would go upwards and the plane would **stall**. If the engines do fail, we would want the plane to go into a 'nose down' position so that it would glide rather than stall. Therefore the usual arrangement in an aircraft is to have the centre of pressure behind the centre of gravity. In this case the aircraft tends to tilt nose downwards, as in Fig. 5.7.

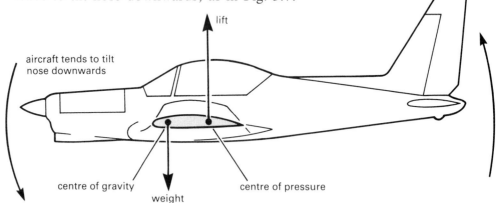

Fig. 5.7 Aircraft tilt tending to nose downward

■ Angle of Attack

When an aircraft changes the angle of the aerofoil surfaces to the flow of air, it is said to change its **angle of attack**. Each particular wing has an imaginary line called the **chord** line. The angle this line makes with the airflow is called the **angle of attack** (see Fig. 5.8).

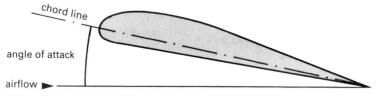

Fig. 5.8 The angle of attack of an aerofoil

The aircraft can increase its lift by increasing its angle of attack (Fig. 5.9). This is only true up to a certain limit. At a particular angle, very often about 15°, the airflow over the wing becomes turbulent and the wing loses lift. Turbulence occurs when the streamline pattern (e.g. Fig. 5.9) breaks down, and then Bernoulli's principle no longer holds true. The result of this loss of lift is a stall, where the aircraft accelerates downwards, out of control, i.e. it falls rather than flies. If the aircraft is at too low an altitude and there is no time to get its nose down to decrease the

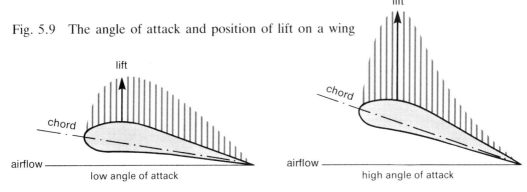

Fig. 5.9 The angle of attack and position of lift on a wing

angle of attack and make it fly again, it will crash! Stall-avoidance is a high priority for both designers and pilots.

Another effect of altering the angle of attack is to move the centre of pressure of the wing. As the angle of attack is increased, the effective lift force moves forwards and increases. This affects the balance or **trim** of the aircraft.

■ Drag

The airflow around a wing produces both lift and **drag**. Drag is one way in which we 'pay' for the lift. Both the lift and drag forces increase with speed but not proportionally, as both are related to the square of the aircraft's **velocity**, i.e.:

lift is proportional to (velocity)^2;
drag is proportional to (velocity)^2.

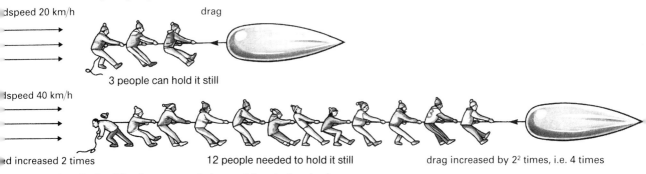

Fig. 5.10 The increase of drag with wind velocity

Drag not only arises around an aircraft's wings but from its whole structure. We can split the drag force into three main types:

(a) form drag;
(b) skin friction;
(c) induced drag.

At high speeds other forms of drag emerge. These will be discussed separately.

☐ Form Drag

Aircraft designers thought at one time that the front area of an aircraft affected the drag. They attempted to reduce the frontal area as much as possible but did not realise that the shape of a body has almost as much effect as its cross-sectional area.

Today we take streamlining for granted on cars, trains and aeroplanes. The reason for designing them in this way is not just to make them look good but to reduce drag, as any reduction of drag means a reduction in fuel consumption. This is particularly important for aircraft due to their very high speeds and their having to carry all of their fuel with them from the start of most trips.

A coin can be streamlined so that its drag in an airflow is reduced to less than 5% of its original value. This reduces the drag to an even lower level than would be achieved by turning the coin end-on to the airflow (Fig. 5.11).

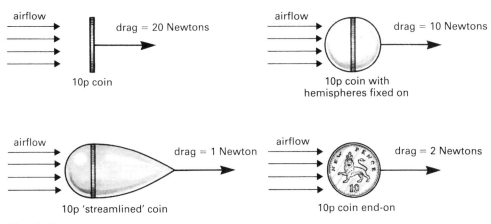

Fig. 5.11 A coin in an airflow

Anything that makes the airflow swirl in a turbulent way will increase the drag. Modern aircraft have as few parts as possible standing proud of the surface as any such parts will produce turbulence.

The biplane was a very manoeuvrable and strong aircraft, but its spars and trusses caused turbulence which increased drag. In an attempt to decrease drag, designers eventually streamlined the tensioning cables and spars. The clean outline of the later monoplane design avoided these problems and this was one reason for it becoming more popular.

☐ Skin friction

If the aircraft shape is streamlined, it may still have a great deal of drag due to the friction between the air molecules and the rough surfaces of the aircraft. High speed aircraft need and therefore have streamlined shapes and highly polished surfaces to avoid both drag and **skin friction**. It was because of skin friction that World War II fighters flew some 20 mph slower when in matt camouflage colours. You can appreciate this if you compare the ease of moving your hand over a smooth bench and over a piece of sandpaper. This explains why ground crews take so much care to keep the fuselage and wings clean on high performance aircraft. Grease from hands acts as a very effective glue for dust which will increase the drag!

The air molecules next to the surface of the aircraft skin are slowed down by the frictional force between themselves and the aircraft's skin so that they are virtually stationary. This is called the **boundary layer** – it may only be a fraction of a millimetre thick but it is almost 'glued' against the surface.

The molecules above the boundary layer are moving slowly and the next layer are moving faster and so on. The tendency of the layers to stick to each other is called the **viscosity** of the material. For example, treacle and tar are very viscous.

For treacle and tar, viscosity decreases with temperature rise, but the viscosity of air increases with temperature rise. Also air viscosity is not pressure sensitive. It does reduce as pressure reduces but not significantly until at very low pressures such as at high altitudes over 10 000 metres. Therefore at 10 000 metres or more skin friction is much lower than at low altitudes.

☐ Induced Drag

The types of drag discussed above are important at high speed but induced drag is most significant at low speeds. The drag here comes from **vortices** or whirlpools of air produced at the wing tips (Fig. 5.12). We can see these vortices if we glue streamers on to the wing tips of an aircraft in a wind tunnel. We would see them spin rapidly and tear themselves to pieces.

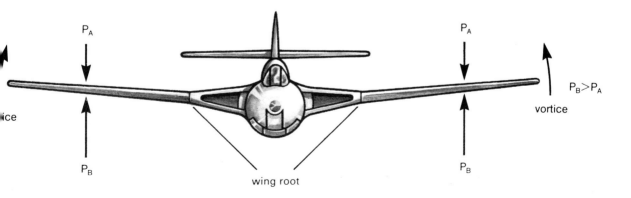

Fig. 5.12   The pressures around an aircraft

It takes work to stir up whirlpools and this work has to be paid for – by extra fuel in a powered aeroplane, or by a steeper gliding angle in a glider.

Air always tries to flow from high to low pressure areas. Air from the higher pressure area under the wing tries to flow around the wing tip to the top surface of the wing where the pressure is less (Fig. 5.13).

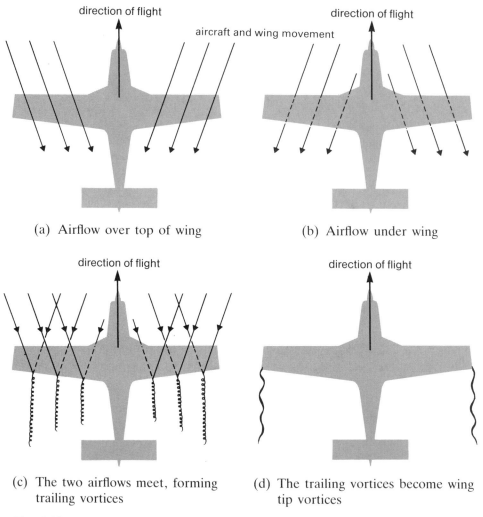

(a) Airflow over top of wing

(b) Airflow under wing

(c) The two airflows meet, forming trailing vortices

(d) The trailing vortices become wing tip vortices

Fig. 5.13 Formation of vortices

In addition, as the pressure alters over the wing surface, air on the bottom surface of the wing tends to move outwards along the surface. Air on the top surface, however, moves inward towards the wing root. This **cross over** of airflow causes turbulence and increases drag at the trailing edge of the wing. The little whirlpools add together to produce a large vortex at the wing tip. The induced drag can be reduced by:

48

(a) increasing the aircraft's airspeed, which gives the air less time to 'cross over' the wing – this is why induced drag is more significant on slow aircraft, such as gliders;

(b) reducing the wing chord, which again gives the airflow less time to 'cross over'.

However, to keep the lifting power, the area of the wing must be made as high as possible. So to reduce induced drag a long thin wing is used. This wing is said to have a high **aspect ratio**.

The long narrow wing with its wing tips wide apart will reduce the amount of air able to flow around the tips and so it will also have a smaller induced drag than the short wing. However, this makes the aircraft difficult to manoeuvre both on the ground and in the air. It is also heavier than a normal sized wing and has special structural problems.

☐ **Other Forms of Drag at High Speeds**

Two other types of drag which occur at high speeds are **shock wave drag** and **interference drag**.

Shock Wave Drag

At high speeds, approaching and above **Mach 1** (the speed of sound), a **bow wave** is caught on the aircraft. This increases the drag considerably. We shall look closer at this later in the book.

Interference Drag

The frontal area of the aircraft is not generally as important as once thought. However, for high speed aircraft, changes in cross-sectional area must be engineered very carefully, or additional drag results. This is particularly noticeable around the wings where the cross-sectional area increases enormously, as Fig. 5.14 illustrates.

Fig. 5.14   Interference drag at high speeds

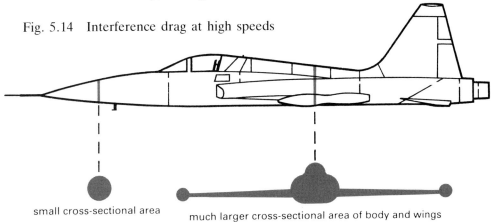

small cross-sectional area   much larger cross-sectional area of body and wings

The change in area can be reduced by reducing the cross-sectional area of the fuselage at the wing roots to compensate for the area of the wings. This is known as **waisting** or **area ruling** (Fig. 5.15).

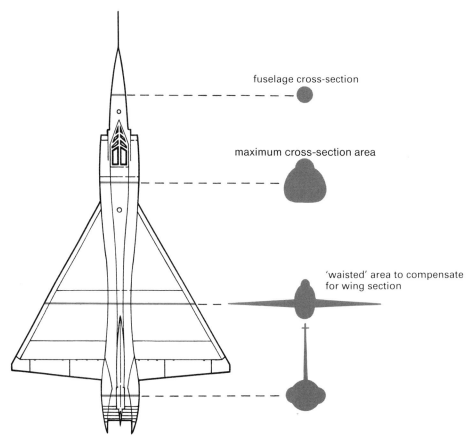

Fig. 5.15 Waisting or area ruling an aircraft reduces (cross-sectional or interference) drag

The Balance of Forces

Dynamically, a force can change the speed, direction, or shape of a body, or any two or all three of these at the same time. Two or more forces acting on a body but not passing through a common point will produce rotation. Also, one force acting on a body but not changing its speed or direction always invokes a second force to react to the first. For example if a book rests on a table, the book is pressing down on the table, but is the table pressing up on the book? The answer must be 'yes' since if the table were not pressing up, the book would move through the table! We can summarise this effect by saying that the action of the book on the table is balanced by the reaction of the table, as illustrated in Fig. 5.16.

Fig. 5.16 Action and reaction

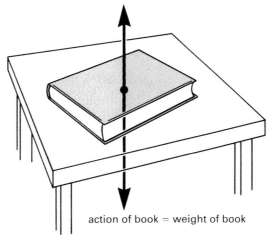

Thus for every action there must be an equal and opposite reaction. This is Newton's third law and applies (strictly speaking) to static situations.

An example of a dynamic situation is the footballer kicking a ball, as in Fig. 5.17. Bear in mind that there are two forces present – the footballer's kick and the weight of the ball.

As far as an aircraft is concerned, it must be strong enough not to deform under the four forces acting on it – weight, lift, thrust and drag.

The job of the lift force is to overcome the weight of the aircraft. The thrust force of the propeller or jet has to overcome the drag forces acting on the aircraft. If all these forces are balanced, i.e. lift equals weight, and thrust equals drag, the aircraft will stay at the same height and move through the air with constant speed. If the lift is more or less than the aircraft's weight, then the aircraft will climb or descend. If the thrust is more or less than the drag, then the aircraft will accelerate or decelerate.

Fig. 5.17 A football kick – a dynamic situation

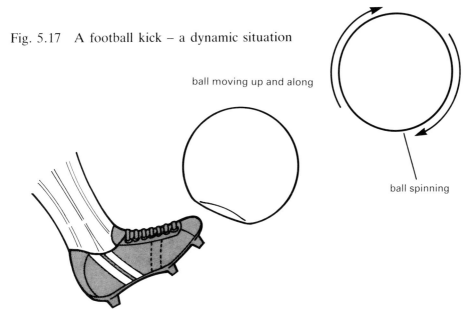

To summarise:

horizontal speed $\begin{Bmatrix} \text{increases} \\ \text{stays steady} \\ \text{decreases} \end{Bmatrix}$ when thrust $\begin{Bmatrix} \text{exceeds} \\ \text{equals} \\ \text{is less than} \end{Bmatrix}$ drag;

altitude $\begin{Bmatrix} \text{increases} \\ \text{stays steady} \\ \text{decreases} \end{Bmatrix}$ when lift $\begin{Bmatrix} \text{exceeds} \\ \text{equals} \\ \text{is less than} \end{Bmatrix}$ weight.

A football does not spin when the line of the kick passes through its centre of gravity. For an object which is free to move, any force acting along a line which does not pass through its centre of gravity gives it rotation as well as translation, i.e. it causes it to rotate as well as to change its position. The further the line of the force is from the centre of gravity, the larger is its rotating effect. The strength of the rotational effect of a force is called its **moment**.

The lift of an aircraft comes from its wings. It will change with the angle of attack and has a line of action through the combined wings' centre of pressure. The weight will act through the centre of gravity.

Weight will change as fuel is consumed, or bombs are dropped. The centre of gravity will move as passengers move and may move as weight changes. If the centre of pressure coincides with the centre of gravity, then the lift force acts through the centre of gravity. The two forces will then have no rotational effect. However, aircraft designers usually ensure that the centre of pressure lies slightly behind the centre of gravity, so that these two forces do exert a small turning movement or **couple**. This ensures that the rotational effect of the forces will turn the aircraft to a safe, slightly 'nose down' position if the aircraft is left to fly by itself. See Fig. 5.18, where these forces have an anticlockwise effect.

Fig. 5.18 The rotation of an aircraft due to the rotational effect of the lift force only

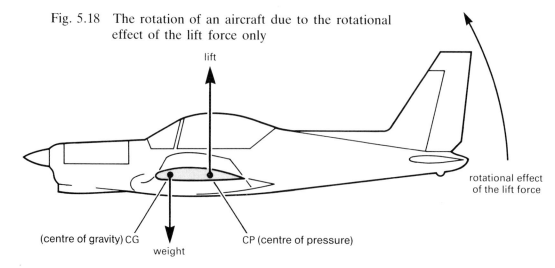

This tendency of the lift force placed in this position to rotate the nose down is normally counteracted by the rotational effect of thrust in the opposite direction about the centre of gravity, i.e. clockwise. For this to happen the line of action of the thrust must lie below the centre of gravity (Fig. 5.19). If additionally the line of action of the drag force lies above the centre of gravity, then it assists the rotational effect of the thrust. If it lies below the centre of gravity, it assists the lift force's rotational effect. In the latter case, the thrust force has to do all the work in counter balancing the 'nose down' effect of the lift and drag. In Fig. 5.19 the drag force is above the centre of gravity and is assisting the thrust's rotational effect.

Fig. 5.19   The balance of forces on an aircraft

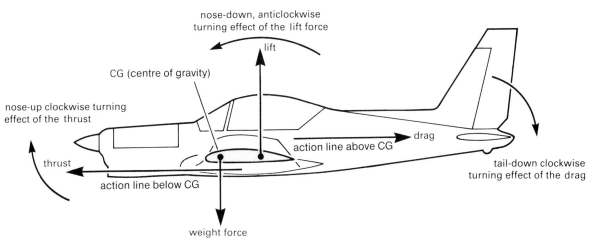

When the rotational effects of lift and weight and of thrust and drag are all balanced, the aircraft will stay horizontal at a particular altitude and steady speed. As we have already discussed, the 'nose down' effect is necessary in case of engine failure when the thrust force ceases, as the aircraft then glides instead of stalling.

☐ **The Tail Plane**

Americans call the tail plane a **stabiliser**. Its job is to provide an additional force. This will be a turning effect that can be used to help balance the rotational effect of the lift force. Since it is a long way from the pivot point of the aircraft, the force produced by the tail plane will have a large turning effect. Therefore a small force produced by the tail plane will have a large effect on the stability of the aircraft (see page 54, Fig. 5.20).

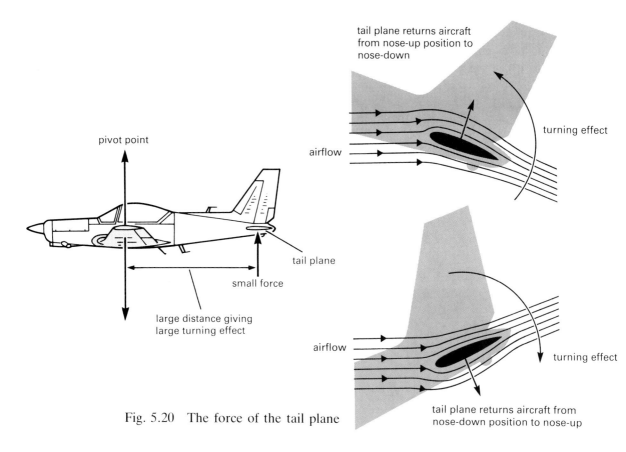

Fig. 5.20 The force of the tail plane

The tail plane is curved equally on both its top and bottom surfaces. When the tail plane has a very small angle of attack, it has no lift. When the tail plane has a negative angle of attack, it experiences a downward force. This rotates the aircraft until it is level. The force is small, but, since it is a long distance away from the other forces, it has a large effect.

There are small adjustable angle sections on the trailing edge of the tail plane. These are the **elevators**. They can be used to adjust the average angle of attack of the tail plane and enable the pilot to control the altitude of the aircraft.

The aircraft must be controllable in yaw and in order to be able to produce forces which have a turning effect in the plane of yaw (horizontal plane), a fin, usually known as the **tailfin**, is constructed at the rear of the aircraft.

6 Power Units

■ Introduction

This chapter is concerned with the ways aircraft are driven, using propellers or different types of jet.

The athlete uses 'spikes' in order to make sure that friction is maintained between her foot and the ground (Fig. 6.1). She pushes the ground in the opposite direction to the one in which she wants to go. Her push is said to be the **action**. The ground's resistance to this action, which causes her to move forwards, is called the **reaction**. Action and reaction are both forces, and are both measured in Newtons. Newton's third law tells us that action and reaction forces are equal and opposite.

Fig. 6.1 An athlete moving – action and reaction

An illustration more related to aircraft propulsion is shown in Fig. 6.2. A person in a small boat on a calm, wind-free day has a supply of heavy stones with him. One at a time, he throws these as hard as he can over the stern of the boat. He has to be properly balanced to do this and must be pushing backwards, towards the front of the boat, each time he throws a stone. This pushes the boat forward each time a stone is thrown. This is what effectively happens with powered aircraft throwing air behind themselves by the action of a propeller or jet engine.

Fig. 6.2 Propulsion

55

■ Aircraft Movement

Newton's third law applies to aircraft as well as to athletes. The aircraft's action is to push air backwards and in so doing it moves itself forward (Fig. 6.3).

Thrust in an aeroplane depends on pushing air backwards to create a reaction, or thrust, forwards. Normally we propel a ship not by throwing stones over the stern but by pushing the surrounding water backwards, using paddle wheels or propellers.

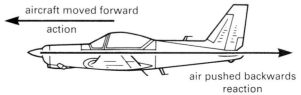

Fig. 6.3 Action and reaction for an aircraft

In an aircraft, air is thrown backwards either by a propeller or by a jet engine. So the force to move the aircraft forwards is provided by pushing air backwards. The air starts off at rest and is accelerated backwards through the jet engines of a jet propelled aircraft or by the propellers of a propeller-driven aircraft.

■ Propeller Thrust

When a propeller is being used, the air is accelerated backwards to form a **slipstream**. In fact the propeller blade is shaped like an aeroplane wing – only it is twisted sharply from the centre and flattens out towards the tips (Fig. 6.4). As the propeller rotates, its blades strike the air at a small angle of attack. This has the effect of driving a stream of air backwards along the axis of rotation of the propeller.

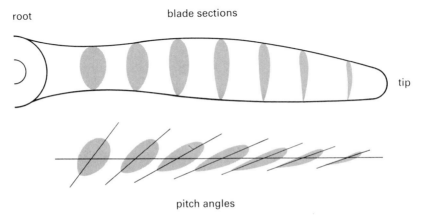

Fig. 6.4 A section of a propeller blade from root to tip

The blade rotates as it moves forwards, so the path that the blade tip follows is a **helix**. A screw thread is an example of a helix, so a propeller is often called an **airscrew** (Fig. 6.5).

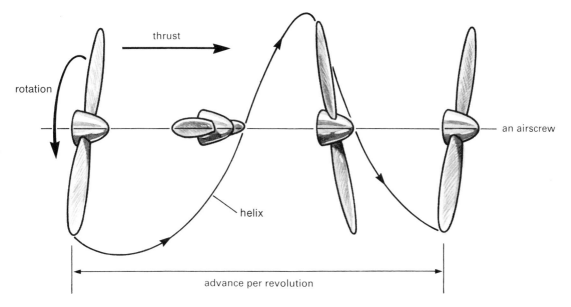

Fig. 6.5 An airscrew

The airscrew does not 'grip' the air like a tyre grips the road. There is always a certain amount of slip. The blade of the propeller is set at an angle to its rotating plane. This is called the **pitch angle**. This is large near the **boss** – the centre of the propeller. The angle decreases gradually towards the tip. This is because although all the parts of the propeller are moving together, the tip is moving faster than the parts near the boss. The blade must have even lift along its length or it would bend and be inefficient. The tip part of the blade at a low angle of attack will give the same lift at a high speed as the boss part of the blade, with a high angle of attack but a low speed.

☐ **Variable Pitch Propellers**

The best average pitch angle for a propeller blade depends on its speed of rotation and the forward speed of the aircraft. To maintain high efficiency over a range of speeds it is necessary to change the propeller pitch angle during flight. Variable pitch propellers were invented in 1932 and are in general use for propeller-driven aircraft.

When an engine fails in a multi-engine plane, it is important that the failed engine's propeller blades have minimum air resistance. With a variable pitch propeller it is possible to turn the blades until they are almost in line with the direction of flight. Then in a stopped position they offer their least possible air resistance. This is called **feathering**.

High speed propellered aircraft often have high landing speeds. In order to have reasonably sized runways it is necessary to slow down the aircraft rapidly on touch-down. With variable pitch propellers it is possible to alter the pitch angle so that the normal rotation of the propeller produces a forward thrust on the air. This gives a backward thrust on the aircraft, which helps to slow it down. This is called **reversed pitch**.

■ Aircraft Engines

An early and efficient propeller was produced by the Wright brothers when developing their biplane. Earlier ideas had been based on the flight of a bird. These aircraft, called **ornithopters** were designed with wings that could move. The designers hoped to produce lift and thrust at the same time but were unsuccessful in achieving flight. This was because the wing motion of a bird is too complicated to be copied mechanically by a movable aircraft wing. Cayley, the 'father of aeronautics', finally decided that it was not practical to copy this method of producing lift and thrust simultaneously.

When early aviators realised that the fixed wing was the most practical solution to flight of any kind, they realised that they were facing a very great problem. They had neither an efficient propeller, nor a light, powerful source of propulsion to drive it! Some aviators hoped to use human muscle power. Others designed machines driven by steam engines. Neither idea was practical at that time because in both cases the power developed was too small when compared to the total weight to be flown.

The first solution to these problems came with the development of the internal combustion engine and the development of an efficient propeller, both by the Wright brothers. They used a four-stroke engine with a reduction gear to drive their own design-efficient propeller. The working sequence of the four-stroke engine is shown in Fig. 6.6.

Fig. 6.6 The working sequence of the four-stroke engine

induction	compression	ignition	exhaust
suck!	squeeze!	bang!	blow!

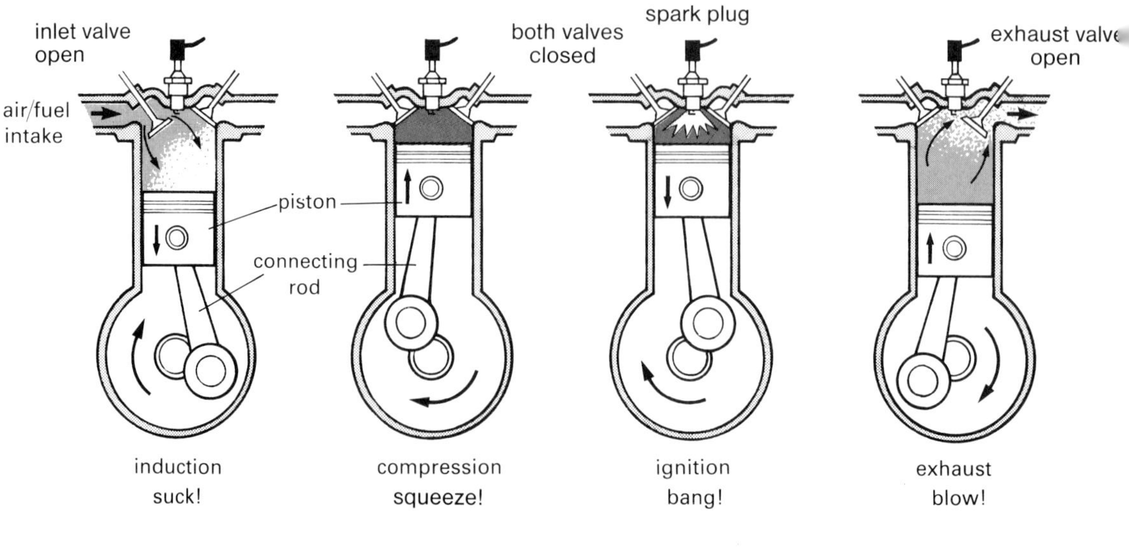

Each cylinder has a **piston** and the pistons in the cylinder go up and down – just like your knees when you ride your bicycle. Each piston is connected to the **crankshaft** by a **connecting rod** (Fig. 6.7). The connecting rods and crankshaft convert the up and down motion of the pistons into the rotary motion of the crankshaft. The crankshaft is connected to the propeller and its rotation makes the propeller rotate. In the early days of aviation the majority of engines were derived from the motor car industry. These were built very solidly, for strength, because weight in a car was not very important. They were made even heavier by their water cooling. Also, they were quite low powered.

One way to reduce weight is to use a coolant other than water for the engine – air for instance. The problem with air-cooled engines is that if the cylinders are arranged in a straight line, the front cylinder will probably be too cool and the rear one will almost certainly be too hot. The solution to this problem was to put all the cylinders in the cold airflow in the form of a **radial** engine (Fig. 6.8). However, this increased the front area of the engine, which increased the drag, slowing the aircraft down. This in turn meant that less cooling air was available to the cylinders. Also, the lowermost cylinders had a tendency to be flooded with oil.

The next development was to fix the crankshaft to the aircraft and allow the cylinders to rotate around it, taking the propeller with them. This heavy rotating mass was now cooled well but it worked like a **gyroscope**. It resisted having the direction of its rotating axis changed. This seriously affected the handling of the aircraft. It made it very difficult to pull out of a steep dive and it restricted manoeuvrability.

From this point on, development concentrated on more powerful in-line liquid-cooled engines, as well as better lubricated, fixed-cylinder radial engines.

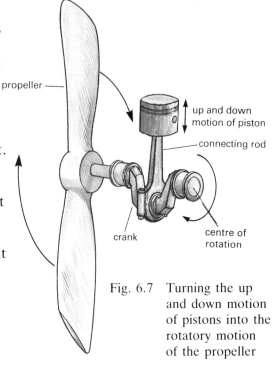

Fig. 6.7 Turning the up and down motion of pistons into the rotatory motion of the propeller

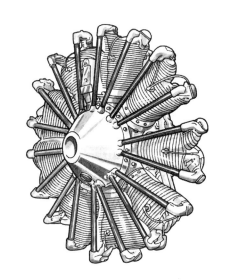

Fig. 6.8 Radially-arranged pistons

☐ High Altitude Flying with Propellered Aircraft

If you use a petrol engine high in the atmosphere, the amount of air that can be drawn into the engine is reduced. This is because the air becomes less **dense** (thinner) as the altitude increases.

The petrol/air mixture passing into the engine can be compressed to compensate for this loss. Power from either the crankshaft or from the exhaust gases can be used to operate a pump or a compressor to increase the density of the mixture going into the engine. Such pumps are called **super-chargers** or **turbo-chargers** depending on the type of pump used.

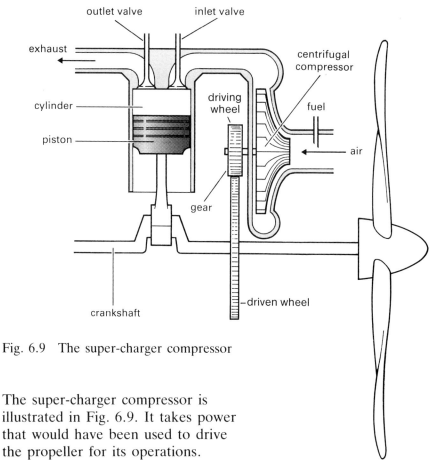

Fig. 6.9 The super-charger compressor

The super-charger compressor is illustrated in Fig. 6.9. It takes power that would have been used to drive the propeller for its operations.

The turbo-charger uses the exhaust gases to turn a turbine, and the motion of the turbine is used to turn a compressor which pushes the petrol/air mixture into the engine. The exhaust gases have nearly a third of the original energy from the fuel so a turbo-charger can make the aircraft more efficient in its use of fuel (Fig. 6.10).

In Fig. 6.11 the engine has been geared down before the power reaches the propeller. This gearing gives the engine the opportunity to increase its power output by increasing the crankshaft speed but at the same time keeping the propeller from going round too fast. The propeller blade tip speed is kept below 0.8 of the speed of sound, or 0.8 Mach, to avoid too much turbulence and loss of efficiency. The power of the engine is taken up by adjusting the propeller blades' angle of attack.

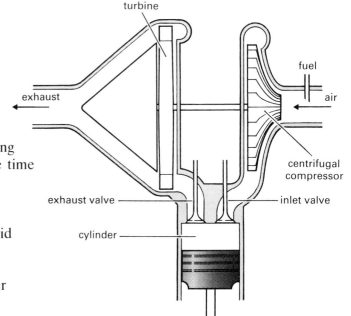

Fig. 6.10 The turbo-charger compressor

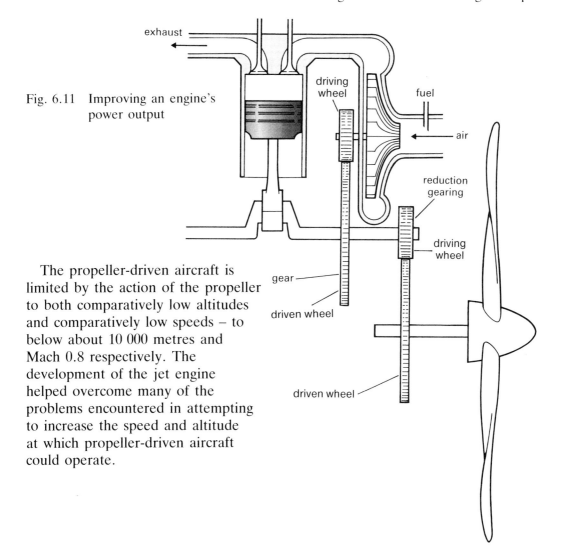

Fig. 6.11 Improving an engine's power output

The propeller-driven aircraft is limited by the action of the propeller to both comparatively low altitudes and comparatively low speeds – to below about 10 000 metres and Mach 0.8 respectively. The development of the jet engine helped overcome many of the problems encountered in attempting to increase the speed and altitude at which propeller-driven aircraft could operate.

Jet Propulsion

The jet engine does not depend upon thrust from a shaped propeller. It produces its thrust by pushing exhaust gases backwards very quickly. It takes in a small volume stream of air and burns paraffin in it. The gases become very hot and expand quickly to make a far greater volume of exhaust gas.

The simplest type of jet engine that is found in use today is illustrated in Fig. 6.12. More properly it should be called a **turbo-jet**. This engine consists of four parts. Taken from the front end, the engine starts with the compressor whose job it is to compress the air taken in at the front of the engine. This compression is required because, just as in the piston engine, efficiency is increased if the fuel is burnt at a high pressure. The compressed air then passes into the combustion chamber, where fuel is sprayed into it. When the engine is running, the fuel ignites spontaneously. This results in sudden increases of temperature and pressure of the gas in this region. The increase in pressure in the gas gives it greater energy than before. The gas then forces its way backwards towards the exhaust, passing the turbine blades on its way. The gas exerts a twisting force on the turbine blades, rotating them. This action drives the compressor, which is on the same shaft. The hot gas then escapes through the exhaust nozzle and in so doing produces a forward thrust on the engine, as predicted by Newton's third law.

This simple type of engine is rather different to the first jet used in civil aircraft. It is found in military aircraft where its small size and light weight are ideal for many applications.

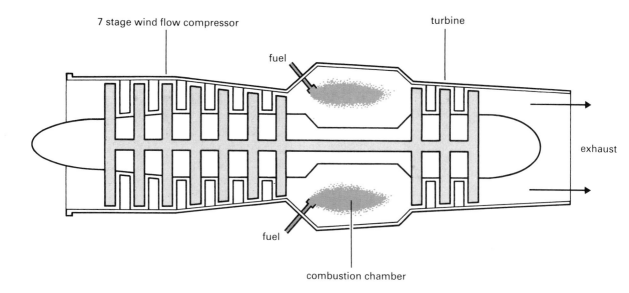

Fig. 6.12 The basic parts of a simple turbo-jet

☐ A Comparison of the Propeller Engine and the Jet Engine

We have now covered enough detail to be able to compare the propeller engine and the jet engine. Figure 6.13 is a detailed comparison of these two types of engine. The most significant difference in their performances is that at low speeds and at take-off the propeller is the more efficient means of producing thrust, while at high speeds and high altitudes the jet is the better.

Turbo-jet	Propeller Engine
Very effective at airspeeds > Mach 1	Very effective at airspeeds < Mach 1
Low fuel consumption at high speeds (> Mach 1)	High fuel consumption at high speeds (> Mach 1)
High 'Ceiling'	Low 'Ceiling'
Smooth continuous motion	Violent intermittent motion
Mechanically simple, reliable, and long lasting	Mechanically complicated, (for 4 stroke engine)
High power/weight ratio	Low power/weight ratio
Power available for driving ancillaries, de-icing, air conditioning etc.	Little power to spare
Comparatively noisy	Comparatively quiet
Inefficient at take-off and low altitudes	Efficient at take-off and low altitudes

Fig. 6.13 Qualities of turbo-jet and propeller engines

The advantage of the propeller at low speeds stems from the fact that a large mass of air is receiving a small acceleration (Fig. 6.14a). The turbo-jet accelerates a much smaller mass of air to a greater velocity, making it much more efficient in sustaining thrust at high speeds (Fig. 6.14b).

Much of the noise generated by an aircraft is the roar which is produced by high velocity air meeting low velocity air. If the difference in the air velocities can be reduced, then the aircraft tends to be quieter.

Fig. 6.14 Comparative airflow diagrams for jet flight (a) and propeller flight (b)

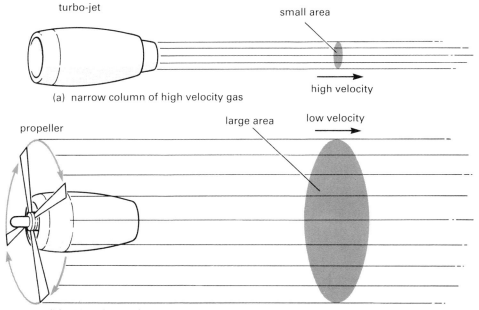

(a) narrow column of high velocity gas

(b) wide column of low velocity gas

☐ The Detail of How a Jet Engine Works

A simple explanation has already been given of how a jet engine produces its thrust. A more detailed explanation now follows.

Figure 6.15 shows the essential parts of an actual **turbo-jet** engine. The air flows in at the front on the left of the illustration, and the hot exhaust gases flow out at the rear, giving the engine its thrust.

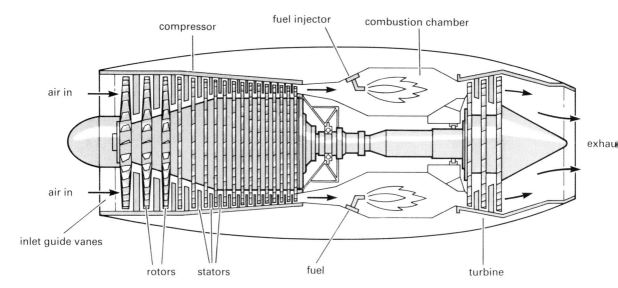

Fig. 6.15   A cross-section of a turbo-jet engine

Air is drawn into the front of the engine by the compressor. The axial compressor shown consists of two sets of blades – the **rotors** and the **stators**. The rotors, which rotate, are set alternately with stators, which do not move. The rotors make the air rotate in their own direction of rotation, thus adding energy to the flow. The combination of rotor and stator converts this rotational velocity into increased pressure. As the air is compressed, its density increases and the area required for the passage of the airflow decreases. This compression causes the temperature of the air to increase further, so that at the rear of the compressor of a modern jet engine the gases may be at 30 times the outside pressure, and at a temperature in excess of 550°C. Air from the compressor passes to the combustion chamber. However, before entering the combustion chamber, the air is slowed down or diffused. This expansion is used to drive the compressor.

The combustor consists of a series of tubes (cans) or an **annulus chamber**, an annulus being a ring-shaped space. Paraffin (kerosene) is sprayed into this chamber and mixed with the air. The fuel/air mixture initially ignited starts to burn and is progressively mixed, or diluted, with unburnt air. Eventually there is complete combustion and an even temperature. About 5% of the inlet pressure is lost in the combustor in order to produce this mixing, but the temperature goes up by several hundred degrees. This temperature rise at roughly constant pressure increases the volume of the gas considerably. The hot, still high pressure gas passes into the turbine. This is similar to the compressor as it consists of stators and rotors. The **rotors** rotate in the same direction as the compressor but they are angled in the opposite direction. This means that the turbine rotors remove energy from the flow of exhaust gases by reducing gas rotation in the direction of rotor blade rotation. The turbine rotors then drive the compressor. In a turbo-jet, the gas leaving the final stages of the turbine still has most of the considerable energy created in the combustion tubes. It is passed through the propelling nozzle to give a high velocity jet which provides the engine's thrust.

In a turbo-prop or turbo-fan engine (detailed in the next section) most of the energy in the combustion tubes is deliberately absorbed by the final turbine, which is specially designed to do this. Then the fan or propeller is driven from the compressor-turbine shaft to provide most of the thrust. The exhaust gases themselves only provide the minor part of the total thrust of this type of engine.

In all gas turbines there is a large amount of energy passing or 'circulating' between the turbines and compressors. With the *RB211* engine at take-off the fan absorbs 25 MW, while the core compressors take 56 MW from the turbines. (Remember, MW stands for a million watts or just over 1300 hp.) Because of this circulation of power it is very important that every component within the core should be as efficient as possible. The piston engine also works on the principle that when a gas is heated either it will expand or its pressure will increase. But here the similarity ends, because in an internal combustion piston engine, combustion takes place in a closed space. The volume remains almost constant whilst the fuel is burnt. This is the volume bounded by the cylinder walls, cylinder head, piston and closed valves. There is therefore an increase in pressure when the fuel is burnt. Then, and only after most of the fuel is burnt, the volume increases as the pressure forces the piston down the cylinder. Of course, the whole cycle does take place very quickly and quite continuously. If you could view it in slow motion, a well-designed internal combustion piston engine would be seen to go through these distinct stages. This is not the case in the turbine where the initial high pressure falls slightly during fuel-air mixture and combustion. This is due to the production of the turbulent flow necessary to produce complete combustion, and the fact that the engine works by allowing the gas to escape at high velocity.

In very simple terms, therefore, a jet engine may be imagined as a fan driven by a windmill with a source of heat in between!

■ The Turbo-prop Engine

Civil aircraft designs combined the new jet engine principle with established propeller practice and produced the **turbo-prop** engine (Fig. 6.16).

The middle section of a turbo-prop engine is identical to the turbo-jet engine already discussed. The difference is that in the final exhaust pipe there is a second turbine. This is connected to a separate shaft that goes right through the engine passing co-axially down the centre of the compressor shaft. This additional shaft drives the propeller at the front of the engine, the drive being through a reduction gearbox for greater propeller efficiency. Propeller engines of this type are lighter weight and smoother running than corresponding piston engines of the same power. Also, however carefully it is balanced, a piston engine uses a to and fro or reciprocating motion that always results in some vibration. This is avoided in a turbo-prop.

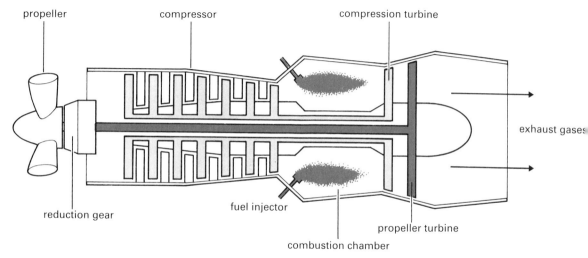

Fig. 6.16 Basic parts of a turbo-prop engine

The engine shown in Fig. 6.16 is called a **twin-spool engine** because the shafts of the compressor and propeller turbines are co-axial and rotate separately. This allows the engine to operate more efficiently, as the best speed for the compressor rotation may not be the best speed for the turbine to take up power to the propeller.

Two of the best examples of the use of these engines are found in the *Britannia* and the Russian *Tupolev Tu 114*, the latter being one of the fastest propeller-driven aircraft ever in regular service. It has a cruising speed of around 800 km/h, but by the early 1980s it was being replaced by later designs – though these were not significantly faster.

The very successful *Viscount* was one of the first aircraft to use a single shaft turbo-prop engine in which the compressor blades and propeller blades were mounted on the same shaft.

For both types of turbo-prop engine, a reduction gearbox is necessary to prevent the airscrew tip exceeding its efficiency speed limit of 0.8 Mach. Figure 6.17 shows the piston-driven engine's efficiency falling off rapidly above 0.6 Mach, with the turbo-prop engine following at 0.8 Mach; whilst the turbo-jet's efficiency is still rising towards its maximum – at speeds well over 1.0 Mach.

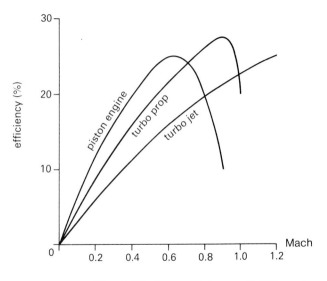

Fig. 6.17 The efficiency of different aircraft engines

■ The Turbo-fan Engine

There are two limitations to the turbo-jet engine:
  (a) its thermal efficiency;
  (b) its mechanical efficiency.

Thermal efficiency can be improved by raising the operating temperature. Mechanical efficiency can be improved by increasing the proportion of kinetic energy of the hot escaping gases converted into useful propulsive energy. The **turbo-fan** engine has reduced exhaust gas velocity, but an increased volume of gas is ejected. This improves the mechanical efficiency, producing the same effect as a much smaller volume being ejected at a higher speed. (This is taken further in the next section.)

The turbo-fan engine was designed to achieve higher efficiencies. As shown in Fig. 6.18, this engine has two compressors. Not all of the

Fig. 6.18 The turbo-fan engine

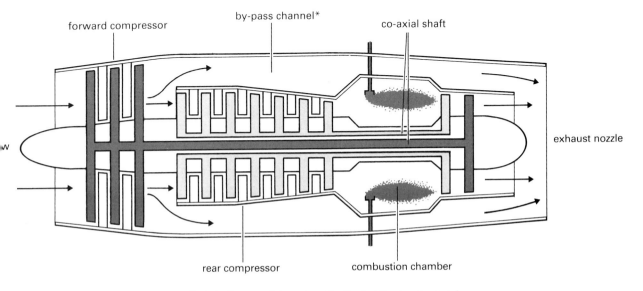

*Often called the by-pass jet engine: the 'fan' at the front proved to be a more popular identification, however. The compressors are on different shafts as their speed of rotation needs to be different because the rear compressor needs to compress the air to a higher pressure.

compressed air from the forward compressor is needed for the rear compressor and combustion. This gas bypasses the combustion chambers and rejoins the hot gas from the combustion chambers in front of the exhaust nozzle. The ensuing mixing expands and cools the gas in the exhaust nozzle. This gives both greater engine efficiency and less noise than a turbo-jet engine.

The turbo-fan engine was first used in the *Trident* and the *BAC1-11*. In fact most presentday airlines use turbo-fan engines.

However, in the future, developments in high speed propellers may bring back the turbo-prop. This could apply especially to aircraft which fly at Mach numbers between 0.7 and 0.8 and are relatively economical in the use of fuel.

☐ The Force Developed by a Jet Engine

In a high performance fighter a relatively small mass of air is given a large acceleration.

In a high bypass ratio engine, such as the *RB211*, a large mass of air is given a small acceleration. Both methods produce a similar result, as can be predicted from Newton's second law. This gives the relationship between force, mass and acceleration as:

force (in Newtons) = mass (in kg) × acceleration (in m/s^2)

For a jet aircraft the mass of air and other gases ejected by the engine in a second is the mass in this formula. The effective acceleration of these gases is given by the difference in velocities between the escaping gases and the aircraft itself. The formula would then be:

force = mass/s of ejected gas × (gas velocity-aircraft velocity)

From this you can see that the same force can be exerted by a large mass of relatively low velocity gas as by a small mass of relatively high velocity gas.

■ Supersonic Engines

It follows then, that supersonic aircraft must use some form of jet engine. The Olympus 593 engine used in *Concorde* is typical of supersonic flight engines (Fig. 6.19). Such engines are more complex than those used for subsonic flight and here we examine the differences. They have two compressors. This means that the gas burns at a far higher temperature, raising the engine's thermal efficiency. In order to prevent the compressor blades breaking, the air entering the first compressor must be travelling at a subsonic speed – even though the aircraft is travelling at supersonic speed. A long air intake is placed at the front of the engine to slow the air down, relative to the aircraft's speed. At the rear of the engine the

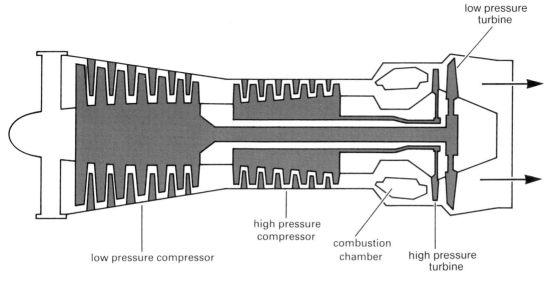

Fig. 6.19 The general layout of the Olympus 593 turbo-jet engine used in *Concorde*

exhaust nozzle opens up as the speed increases so that a larger volume of gas is expelled at a lower relative speed (Fig. 6.20).

At present, designers are developing a supersonic turbo-fan which will be much quieter than the Olympus engine now in use.

Fig. 6.20 Variable intake system and exhaust nozzle

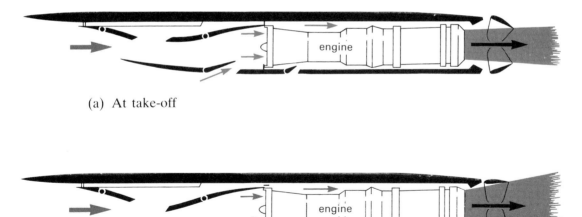

(a) At take-off

(b) At cruising speed

69

Military Jets

At very high speeds, it is necessary to increase the thrust of the aircraft without increasing its frontal area. The exhaust gases coming from the turbo-jet still contain some oxygen and this could be used to burn extra fuel. This system is known as **reheating** or **afterburning** and is very expensive in fuel consumption but does produce significant extra thrust (Fig. 6.21).

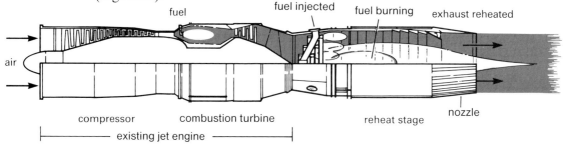

Fig. 6.21 Reheating and afterburning

A World War II development, the **ram jet**, is very simple and efficient. However, it will not work without moving through the air at a high speed, since it has no moving parts to compress the airflow (Fig. 6.22). It therefore has to be launched by a means such as a steam catapult. It has not been used extensively as yet, except for the *V-1* missiles which Germany constructed and used against London towards the end of World War II.

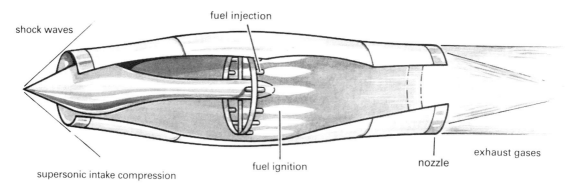

Fig. 6.22 The ram jet

Rocket Propulsion

All of the engines considered so far work by pushing air backwards. The **rocket**, however, is different. It carries its own substances to 'throw out' rearwards to provide forward thrust (Fig. 6.23). So far the use of rockets as the prime propulsion unit on aircraft has been very limited. There was one fighter craft used by Germany during World War II, the *Messerschmidt 163*, but principally since then there have only been the *Bell X-1* and *X-2*, the Lockheed *YF 12A*, and of course the American space shuttle.

The disadvantage of this type of propulsion is that having to carry both the fuel and the oxygen to burn the fuel greatly increases the weight. In a very recent development, code-named HOTOL, engineers have designed a system that can take oxygen from the air. Future space shuttles are likely to use this method for obtaining oxygen on their ascent and descent which will considerably reduce their weight. At present, rocket propulsion is used mainly for travel outside the earth's atmosphere. For both missiles and spacecraft it is so far the only proven system of propulsion.

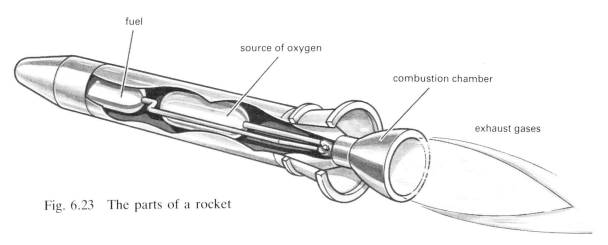

Fig. 6.23 The parts of a rocket

7 Stability and Control

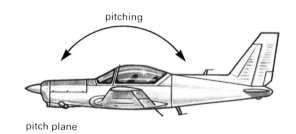

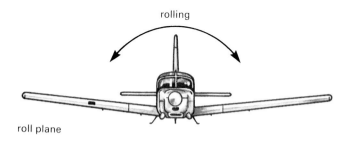

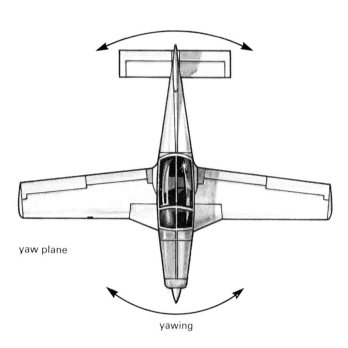

Fig. 7.1 Pitching, rolling and yawing

☐ Aircraft Motion

An aircraft can rotate in three planes:

(a) pitching plane;
(b) rolling plane;
(c) yawing plane.

Pitching is when the aircraft turns to a nose down or nose up position. **Rolling** is when the wings rotate up or down with the fuselage direction unchanged. **Yawing** is when the whole plane turns from side to side whilst moving in a straight line in a horizontal plane. These terms for the movement of an aircraft are illustrated in Fig. 7.1.

An aircraft has to be kept stable in each of the planes.

■ Aircraft Stability

If an aircraft experiences a slight disturbance, it should normally return to its path of flight without any action by the pilot. Any aircraft that does this is said to be **stable** and any aircraft that does not do this is said to be **unstable**. Obviously there are various degrees of stability. In many cases we do not want the aircraft to be too stable so that it can be manoeuvred easily. This is especially true of fighter aircraft.

Stability can be best explained by considering the shape of a 'dunce's cap' (Fig. 7.2). This is a cone. We need a cone that is solid and made of the same material all the way through. Then its centre of gravity will be one-third of its height from its base (along the central axis from the centre of its base to the top of the cone).

If you are very clever, you can balance the cone on its tip – at least for a short time! But you can only keep it there because you keep adjusting the position of the tip of the cone. If you did not, the moment the cone came out of the perfectly vertical position (maybe only a thousandth of a second after you let go of it!), it would start to fall over. Then, unless you interfere, i.e. begin to adjust the position of the tip of the cone again, it must keep falling. That balancing position would be called **unstable equilibrium**, equilibrium being the scientific word for balance.

If you stand the cone on its flat base, you can tip it slightly to one side, and when you let it go, it will come back to rest on its base. This is called **stable equilibrium**. It works up to the position where the cone's centre of gravity is above the point on the circular base which is supporting the cone. Then the cone falls over to lie on its side. The cone lying on its side has its centre of gravity in its lowest possible position (with respect to the surface on which you are considering the movements of the cone). This is called **neutral equilibrium**.

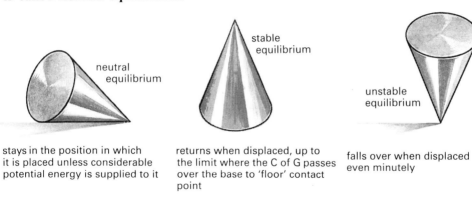

stays in the position in which it is placed unless considerable potential energy is supplied to it

returns when displaced, up to the limit where the C of G passes over the base to 'floor' contact point

falls over when displaced even minutely

Fig. 7.2 The stability of a solid cone

Gravitational stability and equilibrium are all about the position of a body's centre of gravity – how low it can be placed and how well it can be supported. Aircraft stability is very similar – it always concerns the ability to recover the original position after a slight displacement.

■ **Stability in the Pitch Plane**

The aircraft is kept stable in the **pitch plane** by the action of the **tail plane**, together with the four forces of lift, weight, thrust and drag, as discussed at the end of Chapter 5.

These four forces are difficult to balance and tend to be in unstable equilibrium for most aircraft. A parallel situation is that of trying to park a motor bike without a stand or something to lean it against. A side car stabilises the motor bike. Similarly, the tail plane can be adjusted to stabilise the aircraft, i.e. balance the net result of the four forces, weight, lift, thrust and drag, in the pitch plane.

■ Stability in the Roll Plane

When an aircraft is flying horizontally, both its weight and lift forces act vertically and balance each other. If the aircraft rolls over, the weight will still act vertically but the lift will not. It will remain perpendicular to the aircraft's wings. The vertical component of lift will act in opposition to the aircraft's weight. Depending on whether this is greater than, less than, or equal to the aircraft's weight, the aircraft will rise, fall, or remain at the same altitude. The horizontal component of lift will move the aircraft sideways causing a 'wind' to blow over the wings from the other direction to the **sideslip** (Fig. 7.3). This sideslip must be stopped for lateral stability to be maintained. Lateral stability can be achieved using aircraft with at least one of these features:

(a) high wings;
(b) dihedral wings;
(c) swept-back wings.

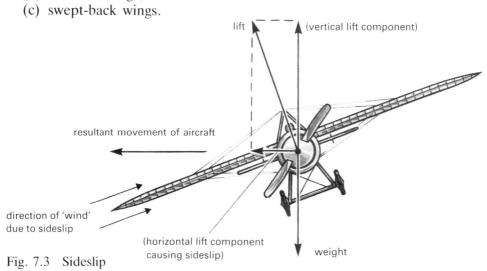

Fig. 7.3   Sideslip

A **high wing** design (Fig. 7.4) places the lift well above the centre of gravity. Early aircraft were designed this way. The centre of gravity tries to find the lowest point possible vertically below the centre of lift. The weight acts like the weight of a pendulum bob. The plane therefore sways from side to side until the movement of the wing through the air curbs the movement.

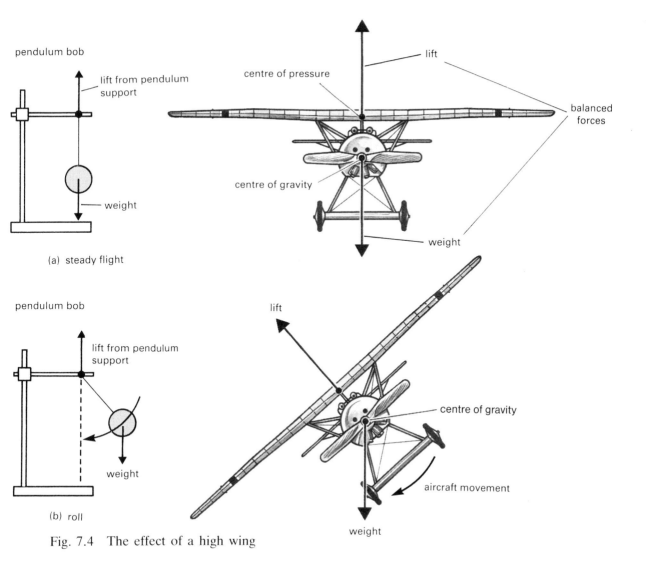

Fig. 7.4 The effect of a high wing

Dihedral wings (Fig. 7.5) are designed so that the wing tips are higher than the wing roots, so that the wings are at an angle to the horizontal. When sideslip occurs, the airflow across the two wings is different. The

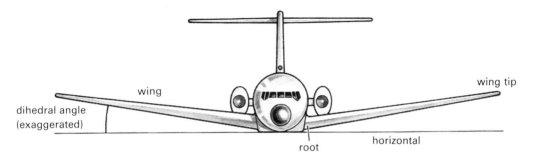

Fig. 7.5 Dihedral wings

75

sideslip 'wind' increases the airflow across the wing that has dropped, giving it greater lift (Fig. 7.6). The other wing experiences less lift because it is protected from the sideslip wind by the fuselage. This results in the dropped wing being brought back to be level with the other wing, correcting the roll. Lift will then balance weight and no further tendency to sideslip will occur.

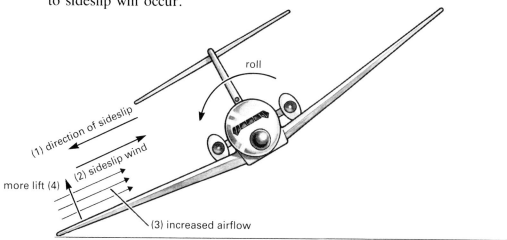

Fig. 7.6 A dihedral wing in roll – effects in sequence 1 to 4

The use of **swept-back wings** (Fig. 7.7) is another way of stabilising disturbances in the roll plane. It is sometimes so effective that an aircraft is too stable for use as a fighter!

The aircraft with swept-back wings has an aerofoil section facing both the main airflow and the airflow coming from the side. So, when the aircraft sideslips (Fig. 7.8), air will flow sideways over the leading wing

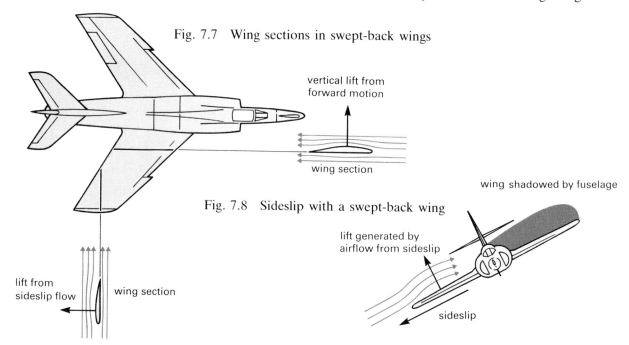

Fig. 7.7 Wing sections in swept-back wings

Fig. 7.8 Sideslip with a swept-back wing

and this wing will experience lift. The opposite wing is shadowed from this airflow and does not experience so much lift. This results in the dropped wing being lifted back to the horizontal. Thus correction occurs as with high or dihedral wings.

These corrections all occur without any action from the pilot. Any aircraft for which this happens is said to have **lateral stability**.

■ Stability in the Yaw Plane

If an aeroplane is suddenly deflected from pointing in its direction of flight, it will still try to continue in its previous direction (Newton's second law) even though it is pointing in a new direction. This will push air over the sides of the aircraft – the **keel area** (Fig. 7.9). The centre of gravity acts as a pivot point for the aircraft and the aircraft acts like a weathercock on a steeple. Then, providing the keel area behind the centre of gravity is larger than the keel area in front, the aircraft will tend to swing back to its old direction. The largest turning moment in this yaw plane comes from the rudder or fin at the rear of the aeroplane. The fuselage itself contributes only a small amount to the restoring force, which is why the rudder or fin is so important.

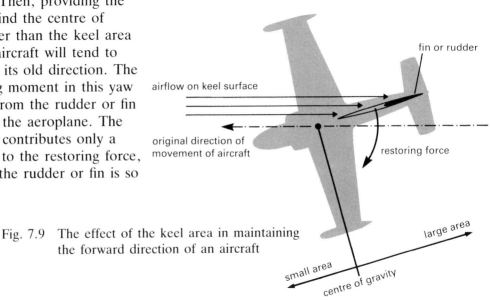

Fig. 7.9 The effect of the keel area in maintaining the forward direction of an aircraft

□ Behaviour of Real Aircraft

We have dealt with stability in the three planes separately because to do anything different would be too complex. You cannot do this in reality. For example, when an aircraft rolls and sideslips, the lateral wind over the wing that rights the plane also acts on the fuselage and tail fin, causing the aircraft to rotate. Therefore, when the aircraft roll has been righted, the aircraft is pointing in a new direction, i.e. *rolling causes yawing*. Similarly, when an aircraft yaws, air flows faster over one wing than the other

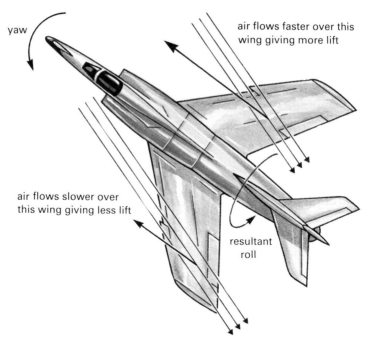

Fig. 7.10 Yawing causing rolling

(Fig. 7.10). This creates more lift on one wing than the other and causes the plane to roll.

This interaction of effects makes designing an aircraft a complex task!

■ Control of an Aircraft

For total control an aircraft has adjustable surfaces that affect its behaviour in all three planes. Aeroplanes do not have a steering wheel as such. They are controlled by means of a control column – often called a **joystick**, and a linked pair of pedals – the **rudder pedals**. The pilot holds the control column (joystick) with his hands and rests his feet on the rudder pedals.

The controls can have some combined effects, which we shall look at later.

Plane	Movement	Controlled by	Worked by
Pitch plane	Nose up or down	Elevator	Control column forward or back
Roll plane	Wings rocking	Ailerons	Control column left or right
Yaw plane	Nose left or right	Rudder	Rudder pedals right or left

■ Control in the Pitch Plane

The **elevators** are the surfaces that provide control in the pitch plane.

The control column is pulled towards the pilot to rotate the elevator upwards. This will increase the airflow's downward force on the tail plane (Fig. 7.11) and make the nose of the aircraft point upwards. It also increases the angle of attack of the wing which increases the lift on the wings. The aircraft then begins to fly upwards at an increasing angle to the horizontal. The control unit is operated by a mechanical, hydraulic, or electrically-powered system. On some aircraft, like the *Trident*, the whole tail plane moves. On delta wing aircraft, which do not have a tail plane, ailerons or control flaps along the trailing edge of each wing produce the same 'nose up' or 'nose down' rotation.

Fig. 7.11 The effect of elevators

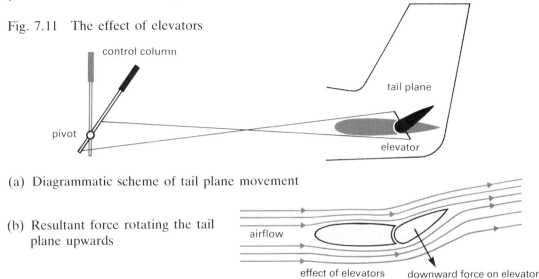

(a) Diagrammatic scheme of tail plane movement

(b) Resultant force rotating the tail plane upwards

(c) Tail planes in normal position

■ Control in the Yaw Plane

The tail fin or **rudder** acts in many ways like the rudder in a ship (Fig. 7.12). It is placed on the tail fin at right angles to the elevator, and is operated by the rudder pedals. In some designs the whole rudder moves, and in others only the trailing part of it moves. Pushing the left

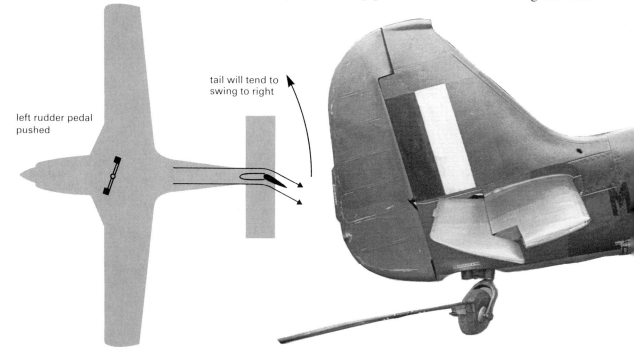

Fig. 7.12   (a) The action of the rudder   (b) The rudder system on an aircraft

rudder pedal causes the rudder to turn left and the aircraft to yaw, nose to the left and tail to the right. This action will turn the nose of the aircraft to point in the new direction. However, the aircraft will tend to slide, rather like a car with smooth tyres on ice (Fig. 7.13). But an aeroplane experiences less friction from the air than car wheels do from the road, and cannot change direction easily by simply altering the alignment of the rudder.

A car will go round a corner with less possibility of skidding if the track is banked. In a similar way an aeroplane will go in a much 'tighter' turning circle if it is **banked** in the direction of the intended turn. In order to bank, the aircraft must be made to roll. This is achieved through the use of the ailerons, as is explained in the next section.

Fig. 7.13   An aircraft attempting to turn with rudder only

Control in the Roll Plane

This is carried out by control surfaces called **ailerons** – movable surfaces set into the trailing edge of the wings. These are set to move in opposite directions in the left and right wings. If the left aileron is turned upwards, the right aileron is turned downwards. The ailerons are operated by the pilot moving the control column to the left or right. Moving the column to the left depresses the right aileron and raises the left aileron, rolling the aircraft to the left (Fig. 7.14). Moving the column to the right raises the right aileron and depresses the left aileron, rolling the aircraft to the right (Fig. 7.15).

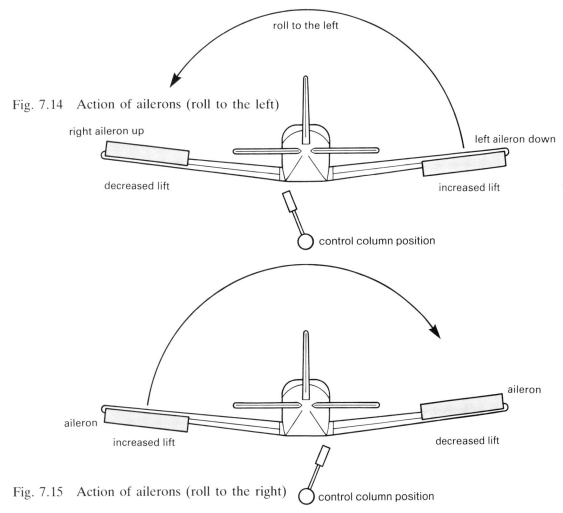

Fig. 7.14 Action of ailerons (roll to the left)

Fig. 7.15 Action of ailerons (roll to the right)

When a pilot wishes to make a turn, it is necessary to roll the aircraft a little as well as to apply the rudder in the desired direction. It is similar to riding a bicycle. You 'lean' into the bends. The turn must be balanced, as on a bike, with the weight acting along the normal axis. The rudder is used to balance the turn.

When the aircraft rolls (banks), a component of the lift force (H) now acts in a horizontal direction, as well as there being a component (V) in the vertical direction (Fig. 7.16). Carrying on with the maneouvre started above, the horizontal component of the lift force pulls the aircraft in towards the centre of the turning circle, preventing the slip. So to achieve a turn the aircraft must roll. But the rudder must be used at the same time to balance the turn, to prevent the aircraft slipping to the centre of the turn or skidding out of the turn.

So with the rudder alone the plane yaws but does not turn. With the ailerons alone the aircraft rolls and will turn in too tight a circle. A mixture of the two produces a balanced turn.

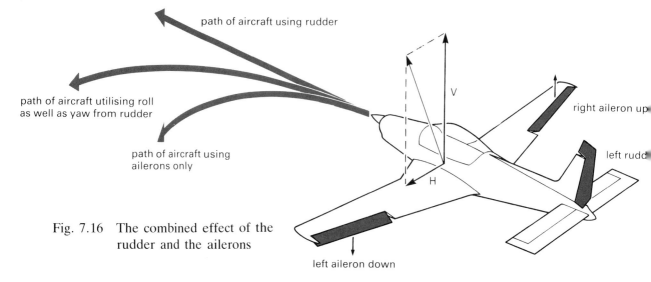

Fig. 7.16 The combined effect of the rudder and the ailerons

☐ **Other Controls**

Flaps, Spoilers, Slots and Slats

All of these are designed to increase the lifting power of the wing at low speeds, mainly when an aircraft is taking off or landing.

Flaps are hinged areas on the trailing edge of the wing. Unlike ailerons they work together in the same direction or in unison. Their effect is to increase the **camber** or curvature of the wing. Increasing the camber of a wing produces an increase in lift. When flaps are used, there will also be a noticeable increase in drag, which will slow the aircraft down (Fig. 7.17).

Spoilers are plates which can be raised at various angles from the upper surfaces of the wings (Fig. 7.18). When a spoiler is raised, the airflow over

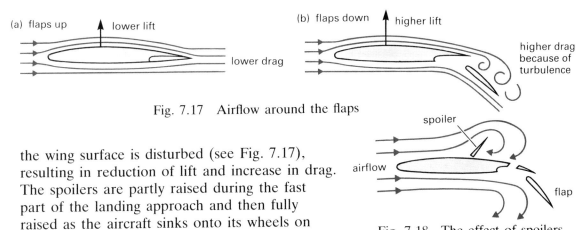

Fig. 7.17 Airflow around the flaps

Fig. 7.18 The effect of spoilers on airflow

the wing surface is disturbed (see Fig. 7.17), resulting in reduction of lift and increase in drag. The spoilers are partly raised during the fast part of the landing approach and then fully raised as the aircraft sinks onto its wheels on landing. This destroys the lift from the wings, putting the aircraft firmly onto its undercarriage and enabling the landing wheels to grip the ground. The fully-raised spoiler has an air brake effect which helps reduce the landing run along the runway. During flight, spoilers on one wing can be raised independently of those on the other wing. Therefore, one spoiler, or more, on one wing can be raised slightly to help bank the aircraft. Slots and slats (Fig. 7.19) are movable or fixed sections on the leading edge of the wing. The **slot** is really a gap between a small extra aerofoil – the **slat** – and the wing. The air travelling through the slot prevents the airflow breaking down into turbulence and causing a stall. This means that a wing fitted with slots and slats can have a greater angle of attack and lift, and therefore a lower stalling speed for landings and take-offs. Figure 7.20 shows the airflow without slots and slats, for comparison with Fig. 7.19.

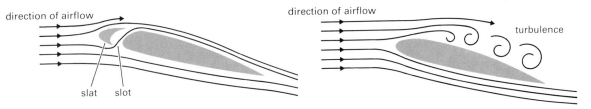

Fig. 7.19 Airflow with slots and slats open   Fig. 7.20 Airflow with slots closed

Trim Tabs

These are small movable sections of the control surfaces and are operated by turning a handle. They are designed to correct any small out of balance forces due to an imbalance of load or the using up of fuel.

In small passenger aircraft the passengers can affect the trim by going aft to the toilet! If all the passengers move, the pilot has to adjust the **trim tabs** quickly. Once the trim tabs are adjusted, the pilot ought to be able to fly the aircraft without putting any constant force on the controls.

8 The Helicopter

■ The Rotor Action

In an aeroplane the lift is provided by the wings. In a helicopter the lift is provided by the **rotor**, which is sometimes called a **rotating wing**. The blades of the rotor are aerofoil shaped and are long and slender (high aspect ratio). The number of blades may vary with the design of helicopter. For example, a large helicopter will require more blades in order to reduce the loading on each blade.

The rotation of the main rotor blade produces a reaction just as the push of the athlete in Fig. 6.1 causes a reaction from the ground of the ground thrust. In other words, Newton's third law applies to rotational motion as well as linear motion. This means that the helicopter fuselage tries to spin in the opposite direction to the rotor blades. To counteract this a small vertical rotor is attached to the rear, as seen in Fig. 8.1. This also provides directional control. The helicopter is controlled by changing the pitch angle of the main rotor. The pitch of the blades can change collectively (**collective pitch**) or be controlled individually according to the position of the blade relative to the body of the helicopter (**cyclic pitch**).

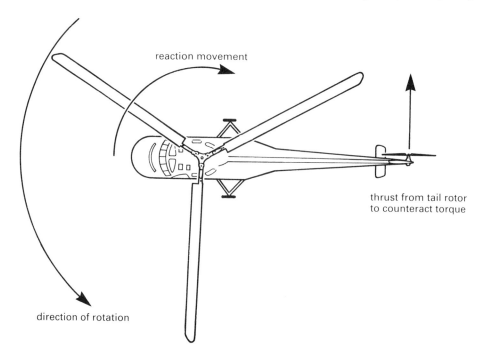

Fig. 8.1 The helicopter

■ Forces Acting on the Rotor

When the rotor blades spin, there are two dynamic forces which act upon them; the **lift** which causes them to go up, and the **centrifugal force** which forces them outwards forming the **coning angle**. The coning angle is the angle between the horizontal plane normal to the axis of solution and the blade axis (see Fig. 8.2). The blades must be attached very securely to the rotor hub! However, to allow them to reach a vertical equilibrium condition, the blades are hinged appropriately at their point of attachment to the hub.

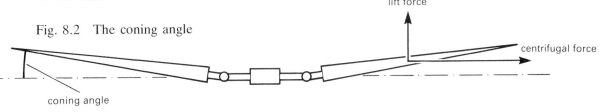

Fig. 8.2 The coning angle

Actually the hinges fitted to the rotor blades allow them freedom of movement in three different directions, as described below. Some modern helicopters are fitted with **rigid rotors**, so called because they are made without drag and flapping hinges, but the blades are strong and flexible enough to take up these movements without damage. However, such rotors would still have pitch-change hinges.

■ The Flapping Hinge

The **flapping hinge** allows the blade to flex up and down because of the variation of lift of the blade as it rotates (Fig. 8.3), thus causing a variation in the coning angle. This is known as the **flapping angle**.

Fig. 8.3 The flapping angle

■ The Drag Hinge

The **drag hinge** allows flexibility back and forward in the direction of rotation (Fig. 8.4). Combining the forward speed of the helicopter and the rotatory speed of the blades, the blade on the advancing side of the rotor speeds up, while the blade on the retreating side slows down. If there were no drag hinges, the blades would bend. The variation in blade position caused by the change in drag between moving forward with the aircraft and moving back is known as the **drag angle**.

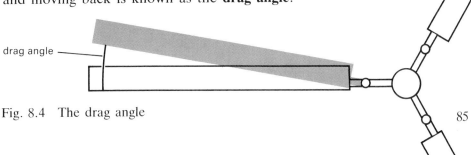

Fig. 8.4 The drag angle

■ Pitch Change

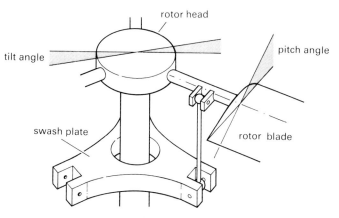

Fig. 8.5 The forward tilt

The third hinge is that which allows a change in pitch of the rotor. As stated previously, the pitch can be changed in two ways. Cyclic pitch is controlled by the swashplate of the rotor head which allows the pitch of the individual blades to vary as they rotate (Fig. 8.5). When the pilot wishes to fly forward, the swashplate is tilted forward. As each rotor blade approaches the forward position (towards the direction of flight) of its cycle, its pitch decreases, the blade lift is reduced and its flight path descends. As the blade rotates to the rear, the pitch is increased, the blade lift is increased, and the flight path ascends.

The net effect is to tilt the whole rotor disc forward to the desired angle. The resultant of all the lifting forces, the total lift vector, is tilted forward and a forward thrust component is given to the helicopter. This is illustrated in Fig. 8.6. L_v is the vertical component and L_H is the horizontal component.

When the lift forces' vertical component, L_v, equals the weight, there will be level flight. When the lift forces' vertical component is greater than or less than the weight, the helicopter will rise or fall respectively. In all cases the forward moving force is provided by the lift forces' horizontal component, L_H.

When the pitch of all blades is changed together by an equal amount, this is called a **collective pitch change**. If collective pitch is increased, the helicopter will climb, and if decreased, it will descend. The collective pitch is changed in conjunction with the engine power settings. This will allow the helicopter to take off, hover, climb and descend.

Fig. 8.6 The forward movement

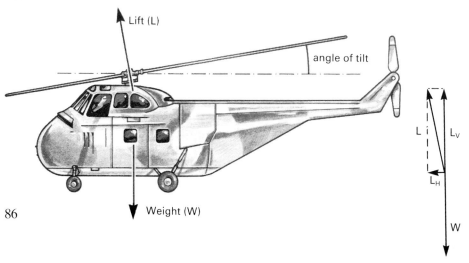

9 Aircraft Instruments

■ The Need for Flight Instruments

In the early days of powered flight, pilots navigated by making use of landmarks or by following rivers or even railway lines. When Charles Lindenberg crossed the Atlantic in 1927, he checked his direction of travel by gliding over a fishing boat and asking which way Ireland was!

Why can't pilots point their planes in the right direction and fly in a straight line? Things are much more complicated. While a plane is flying, it is constantly being blown off course by the wind, its heading may be obscured by cloud or the pilot may be flying in total darkness. Figure 9.1 illustrates in a simplified way how wind alone could cause a pilot to be wildly off course.

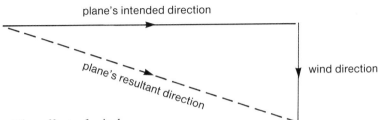

Fig. 9.1 The effect of wind

Even using a good quality magnetic compass, accurate to within 1°, the plane could well be off course by as much as 1 kilometre for every 60 or 70 kilometres of forward travel. It is clear that long-distance flights were very much hit and miss affairs until dependable flight instruments were developed.

Some of these instruments give basic flight information, but some combine information from several sources for the greater convenience of the pilot. Usually the following six instruments will be found:

1 **a sensitive altimeter**, to show the height of the aircraft above the sea or the ground;
2 **an airspeed indicator**, to show the speed of the aircraft through the air;
3 **an artificial horizon**, to show the aircraft attitude;
4 **a vertical speed indicator**, to show how fast the aircraft is rising or descending;
5 **a direction indicator**, to show the compass direction in which the aircraft is flying;
6 **a turn and slip indicator**, to show the rate at which the aircraft is turning and whether or not it is side-slipping or skidding.

These instruments help the pilot to fly the aircraft, especially under conditions of poor visibility, e.g. in clouds or at night. A typical light aircraft flight panel containing these six instruments is shown in Fig. 9.2.

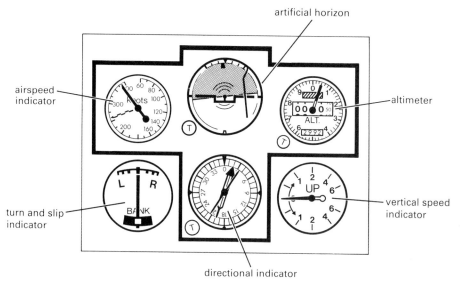

Fig. 9.2 A light aircraft instrument panel

The direction indicator replaces the magnetic compass which cannot be used in a plane because the vibrations from the engines make the needle move about too much, and because such a compass would be affected by changes in the magnetic field outside the aircraft. Several early planes crashed because they flew past hills which were slightly magnetic.

The direction indicator is basically a **gyroscope compass**. A gyroscope is a device rather like a spinning top. It stays pointing in the same direction no matter which way the plane is flying – even upside down. Therefore, once the gyroscope compass has been set in a particular direction, it will continue to give a true reading throughout the flight.

The use of gyroscopes led to the development of the **autopilot**. This device keeps the plane flying in the same direction even if it is blown about by cross winds. Two gyroscopes detect changes in the plane's motion, and then servomotors adjust the controls and put the plane back on its original course.

■ The Air Pressure Instruments

Three of the six basic flight instruments – the altimeter, the airspeed indicator and the vertical speed indicator – depend upon air pressure. A measure of the air pressure is supplied to these instruments from outside the aircraft by a special probe called the **Pitot head**, as illustrated in Fig. 9.3. The dynamic air pressure outside the aircraft is compared with

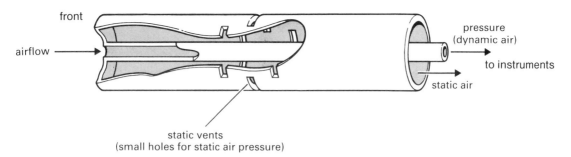

Fig. 9.3 An enlarged view of a Pitot head

the local internal air pressure to operate the instruments. The Pitot head is usually fitted at a low turbulence position on the aircraft; near the wing tip is a commonly used place for this (Fig. 9.4).

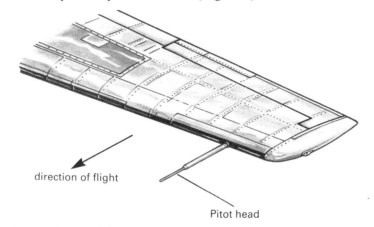

Fig. 9.4 The position of the Pitot head on the wing tip

☐ Inertial Guidance

Supersonic aircraft fly so fast that pilots and navigators do not have time to take the usual readings and bearings – by the time the plane's position has been calculated, it might have flown a further hundred kilometres.

Inertial guidance is based on a computer-controlled map system. Gyroscopes measure changes in the plane's motion and the computer plots the plane's position on a map of the route. No measurements from outside the plane are needed. This system is used in supersonic fighter planes and in *Concorde*.

☐ The Space Shuttle

Perhaps the most advanced use of computer guidance can be seen in the US space shuttle. On its return to earth the shuttle travels down through the atmosphere crossing thousands of miles of air space as a glider. It has no engines to power its return flight. Even a small error could put the shuttle disastrously off course. To avoid this, the navigational calculations are carried out by a 'team' of computers which all work on the same problems. Their results are then passed to a 'chairman' computer which checks to see if they are in agreement. If this is the case, instructions are given on how to bring the shuttle in to land.

Fig. 9.5 The shuttle landing

10 High Speed Flight and Air Traffic Control

■ The Mach Number

As an aircraft's speed increases, it is no longer measured in km/h or mph or knots, but as the **Mach number**. The Mach number is the ratio between the speed of the aircraft and the speed of sound in that locality. This means that an aircraft travelling at the speed of sound is flying at Mach 1.0, an aircraft flying at twice the speed of sound is flying at Mach 2.0, and so on. The speed of sound varies according to the temperature of the air in which it is travelling. As air temperature decreases with height, the speed of sound decreases with height too. At sea level, the speed of sound is about 1200 km/h and decreases with increasing height.

The table below shows the changes that occur with altitude, but precise figures will depend upon the actual temperatures at the different levels.

Height (m)	Approx. Speed of Sound (km/h)
Sea level	1200
3000	1150
6000	1100
12 000	1000

The Mach number is so named in honour of the scientist, Ernst Mach, who investigated the behaviour of high speed projectiles in the nineteenth century.

■ The Sound Barrier

We shall be looking at three different examples of aircraft travelling near or above the speed of sound:

subsonic, below the speed of sound;
transonic or sonic, at the speed of sound;
supersonic, above the speed of sound.

The speed of sound is a very important factor in high speed flight. You may well have heard talk of the **sound barrier**. During the period 1945 to 1955 test pilots, aircraft engineers and designers were trying to make aircraft fly faster than the speed of sound. The table below shows some of the achievements reached.

Date	Plane	Maximum Mach Number
7.11.45	Gloster Meteor	0.79
4. 6.47	Lockheed P80	0.82
25. 8.47	Douglas Skystreak	0.85
15. 9.48	F86 (American)	0.88
19.11.52	F86 (American)	0.92
7. 9.53	Hawker Hunter	0.95
25. 9.53	Vickers Swift	0.96
3.10.53	Douglas Skystreak	≃0.99
29.10.53 to 20. 8.55	F100	≃1.00 to 1.10
10. 3.56	Fairey Delta 2	1.71
16. 5.58	Lockheed F104	2.13
31.10.59	E66 (Russian)	2.25
7. 7.62	E166 (Russian)	2.52
1. 5.65	Lockheed YF12A	3.12

For Mach 1.0 to be exceeded it was necessary to understand what happened at **transonic** and **supersonic** speeds. Once this was grasped, designers were able to start producing the different shapes necessary, and engineers were able to make more powerful engines. This led to a series of improvements in performance and eventually Mach 1.0 was exceeded.

Aircraft at **subsonic** speeds send out disturbances in the air – like ripples from someone rocking a boat on a still pond, as illustrated in Fig. 10.1.

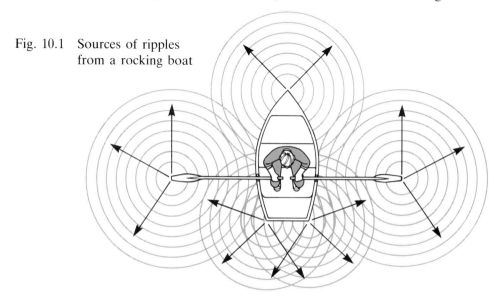

Fig. 10.1 Sources of ripples from a rocking boat

These waves travel at a fixed speed. In water they are the water molecules moving up and down and are known as **transverse** waves. In air the molecules move to and fro in the direction of the travel or propagation of the sound. These waves are known as **longitudinal** or **compression** waves. The speed at which these waves spread out in air is the speed of sound in air.

An aircraft always generates disturbances in the air. These disturbances are sound waves, and they move outwards away from the aircraft at the speed of sound (Fig. 10.2). If the aircraft is moving at less than the speed of sound, the disturbances move faster than the aircraft. They can escape, even from the front of the aircraft, though it will appear to be 'chasing' the waves in front of it. However, information regarding the approach of the aircraft is sent on ahead in the form of these disturbances or waves. As the aircraft flies faster, there will come a speed at which the air flows over part of the wing at Mach 1. This is usually the point on the wing where the curve or camber is the sharpest.

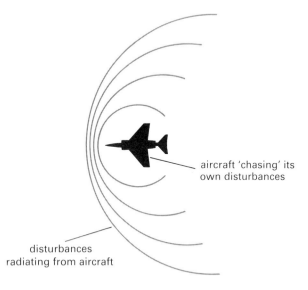

Fig. 10.2   A moving aircraft in relation to its own ripples or disturbances

At this speed the disturbances are not able to move forwards faster than the aircraft itself. Then the air in front of the wing is no longer warned of the aircraft's approach and the disturbances pile up, one upon another. This results in a single large compression wave or **shock wave**. This shock wave disturbs the airflow over the wing, and produces turbulence and eddies behind it, which considerably increase drag (Fig. 10.3). Thus, air will move at the speed of sound over some part of the surface of an aeroplane before the aircraft as a whole reaches the speed of sound. This is likely to be either over the surface of the wings or over a bulge on the

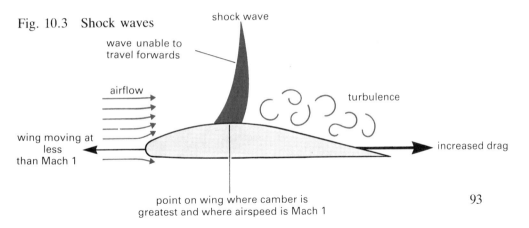

Fig. 10.3   Shock waves

fuselage. This will create a shock wave and this leads to turbulence and increased drag. Increasing the wing camber, or further smoothing round the fuselage, reduces the minimum aircraft speed needed to make the air move over its surface at the speed of sound. The aircraft speed, at which this happens is called the **critical Mach number**. At the critical Mach number, aircraft will experience increased drag and buffeting. The aircraft may even lose lift and experience a **shock stall**.

Suppose a wing is swept back at an angle θ (Fig. 10.4) with air flowing over it parallel to the direction of flight at speed V. The component of speed across the chord V_c, is responsible for the air moving across the wing from the leading edge to the trailing edge. It is therefore this component which will form shock waves when V_c exceeds Mach 1. Figure 10.4 shows that V_c is less than V so that the forward speed of the aircraft will always be greater than the component that creates the shock waves.

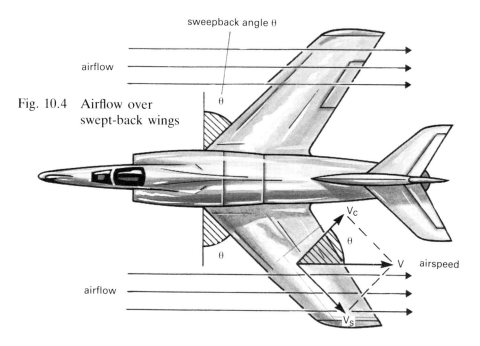

Fig. 10.4 Airflow over swept-back wings

V_s = span component of airspeed, i.e. airspeed along the wing

V_c = chord component of airspeed, i.e. airspeed over wing

A wing of any given cross-section will have a given V_c at which the critical Mach number is reached. For V_c to be kept below this Mach number, θ must be increased in step with V (Fig. 10.5). From the graph shown in Fig. 10.6 we can see that once the aircraft passes through the highest point on the curve, the drag decreases. So, there really is a sound barrier. For the early aviators though, the drag was too great for the wings and engines of aircraft flying and designed at that time.

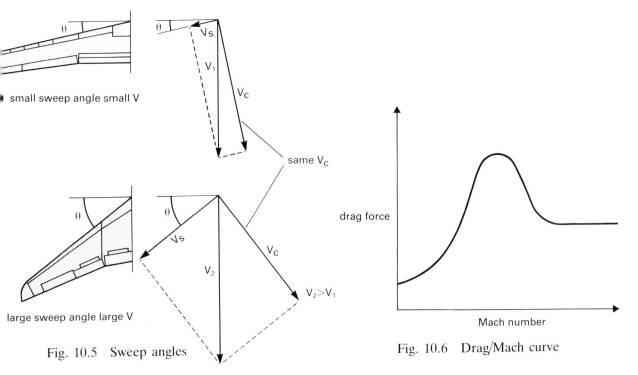

Fig. 10.5 Sweep angles

Fig. 10.6 Drag/Mach curve

If the airspeed increases even more, a second shock wave forms at the front of the aerofoil. Obviously, to enter the world of real supersonic flight, rather than just break through the sound barrier, designers had to find a way to reduce this drag too.

☐ Supersonic Flight

One way of reducing the effect of the shock waves formed when the airflow round an aircraft reaches Mach 1 at any point is to use swept-back wings. It is possible for V to be supersonic while V_c is still subsonic. The faster the aircraft has to fly, the larger θ or the sweep-back has to become for this to remain the case.

The *F111* was designed to have variable sweep wings (Fig. 10.7). They

Fig. 10.7 The variable sweep *F111*

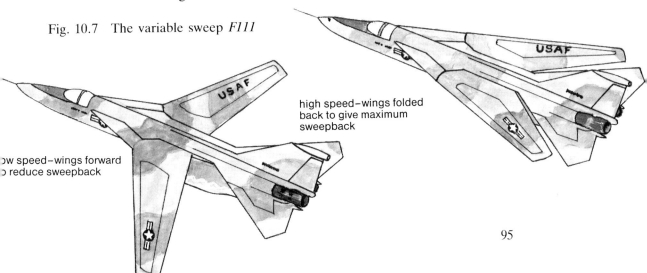

95

are kept forward for subsonic flight and are swept back for supersonic flight. This is because V_c will be too low at low speeds to give sufficient lift. This idea was first proposed by the famous British designer, Barnes Wallis. In the swept-back position, the wings of the *F111* practically join the tail plane.

In a delta-winged aircraft the wing and plane are one which is another method of reducing the effect of shock waves at supersonic speeds. This makes it much easier to have the necessary strength without great thickness, because the main spars are shorter (Fig. 10.8). The name **delta-winged** comes from the similarity in shape of the two wings to the Greek letter delta ($\triangle$). A delta wing achieves the same airflow as a swept-back wing.

Although delta-winged aircraft have a tail plane with a rudder, the ailerons along the trailing edge of the two wings provide all of the real control. When the ailerons on both wings are up, they produce a pitching effect; when the ailerons are up on one side and down on the other they produce a rolling and yawing action at the same time

Fig. 10.8 Wing structure

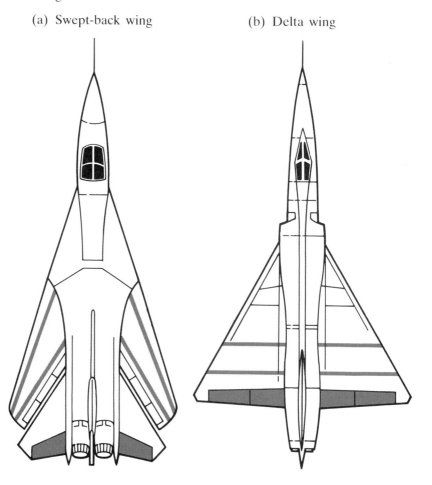

(a) Swept-back wing   (b) Delta wing

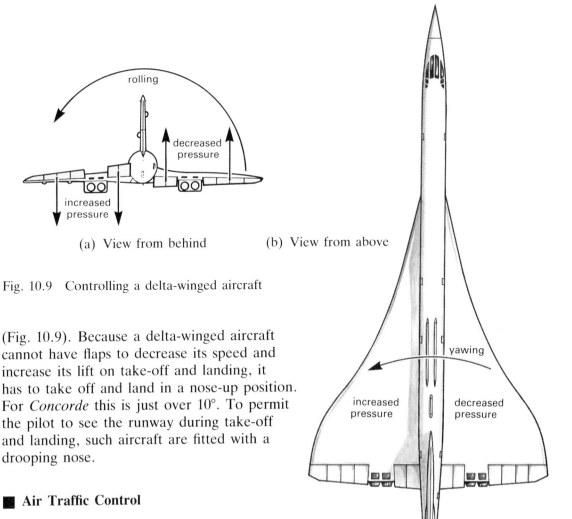

Fig. 10.9 Controlling a delta-winged aircraft

(a) View from behind (b) View from above

(Fig. 10.9). Because a delta-winged aircraft cannot have flaps to decrease its speed and increase its lift on take-off and landing, it has to take off and land in a nose-up position. For *Concorde* this is just over 10°. To permit the pilot to see the runway during take-off and landing, such aircraft are fitted with a drooping nose.

Air Traffic Control

In the early days of aviation there were so few aircraft that the danger of collision in the air was remote. Nowadays there are so many more aircraft flying, and at much higher speeds, that this is no longer the case.

Two aircraft can be approaching one another at a combined speed of up to 1800 km/h, so aircraft need to be controlled carefully in order to avoid accidents!

This is achieved by advising or **controlling** the aircraft from the ground when the weather is bad or when the aircraft flies through certain airspace. Control from the ground means that the pilot is in radio contact with the air traffic control organisation. This uses **radar** and radio aids to follow or monitor the progress of aircraft. The air traffic controller can then direct the movement of aircraft to avoid the risk of collisions, etc.

The whole of the United Kingdom is divided into two flight information regions. Each has an air traffic control centre, with many smaller air traffic control units situated at airports and airfields.

■ Controlled Airspace

It is essential to be able to monitor all aircraft flights as they approach regions where there are higher risks of collision, such as approaches to airports and the regions between these which are called **controlled airspaces**. This monitoring must obviously be done from the ground. This control is even more important when visibility is reduced due to bad weather conditions. However, because of the high speeds of modern aircraft, it is nearly as important under normal conditions to extend the range at which aircraft can be identified and their position, direction and speed monitored. This requires extensive radar systems.

The areas in which this control applies are known as **control zones**, **controlled airspace** and **air corridors** (Fig. 10.10), and examples of these can be found all over the world, linking airports of all types to the major air routes across continents or oceans. In the United Kingdom major cities have airports with local control over all aircraft movements and these airports are linked by air corridors which come under the control of the

Fig. 10.10 Airways and controlled airspace in the UK

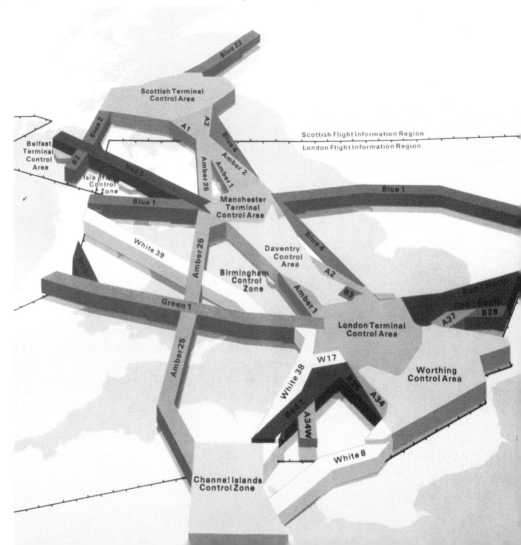

Civil Aviation Authority. The corridors extend from about 1000 metres up to 12 000 metres in altitude and are 10 kilometres wide, except in areas where routes may intersect due to the number of airports and routes.

■ Radar Control

Radar is used to enable an air traffic controller to control an aircraft that he cannot actually see either due to its distance away or to poor visibility. A complex radar system gives information on the identification, position and height of the aircraft.

The radar apparatus sends out a pulse similar to a radio signal. An object in the path of the signal will reflect part of the signal back (Fig. 10.11). The radar will receive this reflected signal, measure the time interval, and convert it into an **echo** on a **cathode ray tube** (similar to a television screen). From the echo, the controller can interpret the height and range of the aircraft and how close it is to other aircraft.

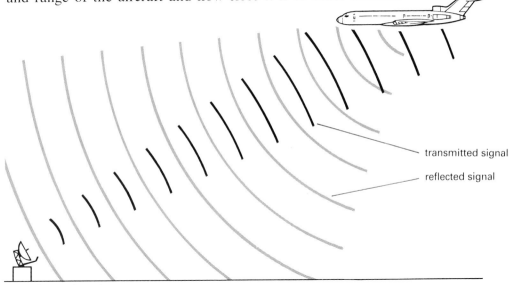

radar transmitter and receiver

Fig. 10.11 Operation of radar

More detailed information can be presented to the controller if the aircraft carries a special type of transmitter called an **air traffic control transponder**. This is triggered by ground radar and then transmits coded signals to the ground. These signals can be displayed as a written presentation of information, such as the aircraft identification and height, and even words like 'hijack' and 'SOS' when the pilot dials the appropriate code.

There are many more radio aids for the control of air traffic. The objective is always to give the pilot and controller information to help to ensure that the aircraft can fly safely to its destination.

11 Aircraft in the Service of Man

■ **Introduction**

Aircraft, like televisions and motor cars, are regarded as part of the 'twentieth-century scene'. We use aircraft to carry passengers and cargo, as vehicles for pleasure, as weapons in war, in agriculture to spray crops, and so on. People have been flying in powered aircraft for only about 80 years. However, in that time aircraft have developed from the flimsy machines built by a handful of enthusiasts to ones capable of supersonic flight such as *Concorde*. But *Concorde* required the resources and technology of two countries for its development, being a joint Anglo-French enterprise.

Aircraft design presents a technological problem for the buyer, the designer, and the people that eventually use the aircraft either as staff or passengers. The buyer – the person or group who specifies what they want from an aircraft – has to decide on the best balance for them, bearing in mind everything that is possible from modern aircraft. They must choose what they would like from their own aircraft, if they can afford it! They must finally decide on a specification that falls within their cost limits.

It may be that the buyer looks for an established design. Here the design team has produced a design that they feel will satisfy a large number of possible customers. The user has to operate the aircraft efficiently and safely. These people have to be trained and their work has to be checked. With an established design this could be easier than with a new design.

Society has a love-hate relationship with aircraft. On the one hand, aircraft provide a means of reasonably cheap, long-distance travel, and a means of defence. On the other, aircraft create noise which damages the lifestyle of people living in the airpath, particularly near airfields and airports, and aircraft consume a lot of fuel.

Liquid hydro-carbon fuel is derived from oil, one of the rapidly running-out fossil fuels. Long after its use for heating and electricity or road transport has ceased, oil will probably be reserved for flying. But this should be well into the twenty-first century at foreseen rates of use, and by that time **synthetic liquid fuel** is likely to exist.

The technological problems of aircraft design start with trying to produce an aircraft that will carry out a particular task. This is influenced by many different factors – costs, scientific discoveries, safety, noise, speed, and so on. Problems continue even when the aircraft is in service. Most service problems are minor ones which have to be put right by modification, re-design or maintenance work.

The technological problems of aircraft design began with the very basic problem of building a heavier-than-air machine capable of carrying a person. The early attempts at solving this problem involved copying the birds. This is why early designs looked like birds, and attempted to fly like them. These were unsuccessful since, as we now know, humans are just not designed to develop sufficient muscle power to fly like birds. Man-powered flight had to wait for high technology materials – and then took place in machines which were virtually high aspect ratio monoplane versions of the Wright brothers' planes!

On 23 August 1977 the *Gossamer Condor* was flown round a figure of eight course to win the Kremer £50 000 prize. On 25 April 1979 the *Gossamer Albatross* (Fig. 11.1) flew across the English Channel to win its team the Kremer £100 000 prize. Both were very lightweight aircraft, made largely of synthetic materials, and powered by a pilot who operated foot pedals to drive a pushing propeller.

Fig. 11.1 The *Gossamer Albatross*

The development of lightweight internal combustion engines enabled the early pioneers, such as Wilbur and Orville Wright, to attain man-carrying, powered flight. By 1905, the Wright brothers had produced *Flyer 3* which was a fully practical aeroplane. For the first time an aircraft was able to achieve sustained and fully-controlled flight.

Pioneers in other countries followed the example of the Wright brothers. Competitions were organised which stimulated designs from all over Europe. *The Daily Mail* offered £1000 for the first aviator to cross the English Channel. The prize was as (if not more) valuable in its day

Fig. 11.2 The Bleriot monoplane

as the Kremer £100 000 prize. It was won on 25 July 1909. Louis Bleriot set off from Calais and landed 40 minutes later in a field behind Dover Castle. He had flown the English Channel at an average speed of 58 km/h (see Fig. 11.2).

■ World War I

World War I produced a rapid change in the development and use of aircraft. At the outbreak of war in 1914, aeroplanes were used for spotting the movement of troops and passing back information to artillery batteries. The main British type of aircraft, the *BE2*, was specially designed to be highly stable in flight so that it could virtually fly itself while both pilot and observer scanned the ground, made notes and took photographs.

Fighter aircraft developed from the aircrews shooting at one another, with hand-held weapons. The technological problem of designing and flying a fighter aeroplane involved the consideration of such limitations as:

(a) stability and instability;
(b) strength and weight of the airframe;
(c) mounting and operation of the guns.

If an aeroplane is too stable, it will be slow to respond to the controls. If too unstable, it will be quick to respond but difficult to control. The *Sopwith Camel* was one of the most agile aircraft ever made but it was too difficult to control and killed many inexperienced pilots.

By the end of World War I, there were fast and highly manoeuvrable fighter aircraft like the *SE5* (Fig. 11.3). Other aircraft had also been developed which could carry heavy bomb loads over long distances, and drop them with considerable accuracy.

Fig. 11.3 The *SE5*

■ Development Between the World Wars

After World War I some of the larger bombers took on a different role. Air passenger services were being established in various countries. It was also the beginning of an era which was to see many historic flights. During June of 1919, Captain John Alcock and Lieutenant Arthur Whitten Brown flew non-stop across the Atlantic Ocean in a converted Vickers' *Vimy* bomber (Fig. 11.4). Other long-distance flights soon followed.

Fig. 11.4 The Vickers' *Vimy*

Air routes were being established over Europe, the Middle East and eventually the Far East and Australia. Communication by flying was truly established. The Air Mail service was started and soon became an important part of air transport. Many people were now becoming 'air-minded' and flying clubs were started by enthusiasts. To meet the needs of these flying clubs, new types of aeroplanes had to be designed. These were the light sporting aeroplanes, such as the De Havilland *Moth* (Fig. 11.5).

Fig. 11.5 The De Havilland *Moth*

During the 1920s and 1930s most aircraft were biplanes, especially military aircraft where the stresses in service were expected to be very high. However, there were advances in materials technology, particularly in the development of light aluminium alloys. This led to the use of cantilevered structures in aircraft design. Wing structures were built around I-shaped section beams of these lightweight yet strong materials (Fig. 11.6).

Fig. 11.6 The Short *L17*

Fig. 11.7 The *DC3 Dakota*

Newly-developed strong materials meant that an aeroplane structure could have a lot of strength in its skin. Therefore, aircraft could have large open cargo areas and, eventually, wide fuselages. The *Boeing 247D* and *Douglas DC2* could be regarded as the ancestors of the modern airliner. Both aircraft were twin-engined, all metal, low wing, cantilever monoplanes. They had a stressed skin structure, retractable undercarriage and variable pitch propellers. The variable pitch propeller gave the aircraft the capacity to fly on one engine, and improved take-off and landing performance. The *DC2* was developed into the famous *DC3 Dakota*, many of which are still flying today (Fig. 11.7).

Civil flying continued to increase. In 1936, the Short brothers' *523* 'C' class flying boats became the mainstay of the Imperial Airway service (Fig. 11.8).

Fig. 11.8 A Short flying boat

Fig. 11.9 The *Supermarine S6B*

Much of the new technology used in aircraft was developed and tested during the many competitions and races which were organised at this time. One of the most exciting air races was the Schneider Trophy Race. This was eventually won outright by Great Britain who entered the Vickers' *Supermarine S6B* (Fig. 11.9). This was designed by R J Mitchell and powered by a Rolls Royce engine. The race was won at a then record speed of 340.08 mph. The *Supermarine* was, of course, the forerunner of the famous *Spitfire* (Fig. 11.10).

Fig. 11.10 The Vickers' *Supermarine Spitfire*

Women in Aviation

In 1824 there were 22 women balloon pilots in Europe, but as powered flight developed in the 1900s, women encountered difficulties in entering aviation. This was due in part to the social conventions of the time. Many people believed that women should not even become passengers let alone pilots and Wilbur Wright created a small sensation when in 1908 he made a short flight carrying Mrs Hart O'Berg – the first woman passenger.

In Britain, the early women pilots were usually wealthy such as Lady Heath who became famous for her record-breaking flight from Cape Town to London. She was an active feminist and campaigned for more women to take up flying.

In the USA at this time, several women found the necessary money for flying by taking part in spectacular 'Flying Circuses'. The public were eager to see planes looping the loop and 'wing walkers' who stood on the wings as the pilots performed dangerous stunts. Women pilots attracted more interest than men and this allowed them to earn enough money for serious flying.

Harriet Quimby, a famous demonstration pilot, became the first woman to cross the Channel when in 1912 she flew from England to France. Because of the weather conditions it was one of the most dangerous flights to be carried out at that time. Thick fog meant that Harriet was unable to see the water during the entire flight and had to rely on her compass alone for navigation. Three months later Harriet Quimby was killed at an air-meeting when her passenger stood up unexpectedly and upset the plane's centre of gravity.

The period between World War I and World War II saw a change in people's attitudes towards women. It became generally accepted that they made reliable pilots. At the same time planes became cheaper and safer. Flying clubs became popular and more women were able to take part. Public interest then shifted away from stunt flying and towards setting up new air routes and breaking records for long-distance flights. The public became a little bored with the subsequent achievements of male pilots and much more interest was shown in the achievements of women.

It was this interest which sparked off the career of Amelia Earhart. In 1928 she became the first woman to cross the Atlantic. Although she maintained that she was 'only a piece of baggage' on the flight, the world's press showered her with attention and virtually ignored her male pilot!

Amelia's popularity grew when she began her own flying career and shortly established an altitude record of 5613 metres in 1929. Three years later she performed a solo flight from Newfoundland to Ireland. She used her fame to encourage yet more women to enter aviation.

Perhaps the most famous woman pilot of this period was Amy Johnson. Working as an office secretary, she saved up enough money to join a flying club and qualified both as a pilot and as a ground engineer. She became world famous after her record-breaking solo flight from London to Darwin in Australia in 1930. Before this trip she had never flown further than the flight from London to Hull!

Her journey to Australia proved to be very difficult with several crash landings on the way, but, making full use of her engineering skills, Amy was able to carry out running repairs and with great courage fly on into history. Amy went on to break many records, including the flight from London to Cape Town.

Amy Johnson's career was important because it made the public take women pilots seriously, and in 1934 Helen Richey became the first woman commercial airline pilot.

During World War II women were employed as pilots and crews of transport planes, releasing men to fly in combat machines, and made a major contribution to the war in the air. Amy herself was a transport pilot during the war and when she was killed when her plane crashed in mysterious circumstances in 1941, the country mourned the passing of an outstanding pioneer.

Throughout the war many thousands of women served in the Women's Royal Air Force in every theatre of operations and the barriers of discrimination began to fall and have been falling ever since. The achievements of women in aviation both as pilots and as ground crew and support staff are such that it is no longer remarkable to find women engaged in what was a predominantly male preserve earlier in the century.

Fig. 11.11 Amy Johnson

World War II

The outbreak of World War II resulted in many advances in aeronautics, with new design features coming into production much more rapidly than in peacetime. A completely new concept in fighter aeroplanes was extensively developed during the war. The *Hurricane* and *Spitfire* (Fig. 11.10) had retractable undercarriages, machine guns mounted in the wings, enclosed cockpits and partially-stressed skin structures.

By the end of the war in 1945, the light alloy, stressed skin construction had been confirmed as the best method of building the basic structure of an aeroplane.

Two significant advances, outside those of aircraft design, were those which took place in air traffic control with the use of ground and airborne radar. These two advances played a considerable part in the successful use of aircraft during the war.

Towards the close of the war, the piston engine was nearing the end of its development, one limiting factor being the ability of the propeller to absorb and transmit the power. A new power unit, the turbo-jet engine, was developed in this country by Sir Frank Whittle. Germany was also involved in jet engine development. Both the *Gloster Meteor* (Fig. 11.12a) and the *Messerschmitt Me262* (Fig. 11.12b) saw service at the end of the war.

Fig. 11.12 Two similar design solutions arrived at separately by two countries at war

(a) The *Gloster Meteor*
(b) The *Me262*

Developments After World War II

After World War II, there was a rapid growth in air travel. At first, many of the aircraft used were developed from wartime bombers and transporters, e.g. the *Lancastrian* from the *Lancaster* bomber (Fig. 11.13), the *Halton* from the *Halifax* bomber, the *Avro York*, which used the wings and engines of a *Lancaster* bomber, and the *DC4* airframe with Rolls Royce Merlin engines to make BOAC'S *Argonaut* (Fig. 11.14).

Fig. 11.13 The *Lancastrian*

Fig. 11.14 After the war, Canadair took the *DC4* airframe and installed Rolls Royce Merlin engines; called the *Argonaut* by BOAC, this type was rather noisy

Fig. 11.15 The De Havilland *Comet 1*

The 1950s saw the introduction of the first pure jet passenger service. This service was started in 1952 by BOAC who used the De Havilland *Comet 1* for the London to Johannesburg route (Fig. 11.15). The investigations which followed the *Comet 1* accidents led to a new and much safer breed of aircraft, *Comet IV*. Nowadays the complete airframe of a new aircraft design is tested under conditions which simulate those experienced by an aircraft in service. For example, the fuselage of a pressurised aircraft is placed in a large water tank and tested under pressure, the tests being repeated many thousands of times and exceeding the limits actually likely to be experienced.

Aircraft appeared with swept-back wings and a delta-plan form. The large jets such as the *Boeing 707*, Vickers *VC10* (Fig. 11.16), and *Douglas DC8* were now taking over the Atlantic air routes. Subsequently there was the increased use of wide-bodied jets such as the *Boeing 747* (Fig. 11.17), *Douglas DC10* and *Lockheed Tristar* (Fig. 11.18), each of which can carry several hundred passengers non-stop up to 9000 kilometres (6000 miles).

Fig. 11.16 The *Super VC10*

Fig. 11.17 The *Boeing 747*

Fig. 11.18 The *Lockheed Tristar*

In the Korean War of 1950–3 helicopters were used in great numbers for the first time. Their ability to get into areas of difficult terrain (Figs. 11.19 and 11.20) made them invaluable as rescue and ambulance aircraft.

Fig. 11.19 Troops and supplies brought in by a Westland *Whirlwing*

Fig. 11.20 Air-sea rescue by a Westland *Sea King*

Fig. 11.21 A *Harrier* taking off

The 1960s and 1970s saw major developments in the fields of Vertical Short Take Off and Landing (V/STOL), variable geometry (**swing wing**) aircraft and, of course, supersonic transport.

The *Harrier* (Fig. 11.21) is powered by a Rolls Royce Pegasus engine which has vectored thrust. This means that the thrust can be directed downwards (for take-off and landing) and backwards for normal forward flight.

The variable geometry wing of, for example, the *Panavia Tornado* and the *F111* enables the pilot to vary the degree of sweep back to the speed of the aircraft. Swept-back wings bring a reduction in drag at high speeds but generate little lift at low speeds, e.g. at take-off and landing. Therefore, the pilot can have straight wings for take-off and landing, sweep them partly back for cruising, and sweep them completely back for maximum speed.

The introduction of *Concorde* into airline service has meant that passengers can cross the Atlantic at twice the speed of sound and arrive at a business meeting with very few after-effects of the journey. You could fly to America in the morning, attend a meeting and return home to Britain in time for supper!

Other types of aircraft have proved themselves useful as crop sprayers and a special type of aeroplane has been developed for this role (Fig. 11.22). These **Ag aircraft**, as they are called, have a characteristic shape determined by the need of the pilot to be able to see the ground immediately in front and to the sides. The chemicals are carried in a hopper and pumped to the spray booms which are attached to the trailing edges of the wings.

The fuselage side panels are normally removable so that the structure can be hosed down after a day's work to get rid of the often corrosive chemicals. Spraying aircraft have to be able to carry large weights of chemicals and must be easy to fly whether empty or fully loaded. Because they fly so near the ground and so close to potentially dangerous obstructions, they must also be easy to fly during the few seconds needed to dump their load in an emergency, and give their pilots a good chance of walking away from a crash!

Fig. 11.22 A *Cessna* agricultural aeroplane

Fig. 11.23 The Britten-Norman *Islander*

For many people throughout the world the airstrip has been the link with civilisation. Small rugged aeroplanes have been evolved to meet conditions found in jungle, desert and frozen waste. The aeroplane is ideal for the transport of medical supplies, food, equipment, and people in such remote areas. Doctors can fly to visit patients, whereas previously visits would have been impossible. An excellent example of this type of aircraft is the Britten-Norman *Islander* which is much used in remote places (Fig. 11.23).

Aircraft can be used to fight forest fires. The Canadair *CL–215* (Fig. 11.24) carries a large water tank and 'bombs' the fire with water. This aeroplane can land on water, take on a fresh supply and fly off again, saving valuable time, which is of course vital when attempting to bring fires under control.

Fig. 11.24 The Canadair *CL-215*

Fig. 11.25 The *Rutan Long EZ* with its proud builder-owners

People's enthusiasm for the pleasure and sport of flying has never been greater; we now see more people taking part in private flying. Many people build their own aircraft (Fig. 11.25). Gliding, parachuting, and hang-gliding are all popular developments associated with flying.

Modern aircraft such as *Concorde* (Fig. 11.26) cost so much that they are very likely to be kept in service for a long time. The jet engine, which runs for much longer with less servicing than the piston engine, encourages operators to keep aircraft operating for as long as possible.

Fig. 11.26 *Concorde*

Fuel prices fluctuate, but the expected long term trend is for fuel to become steadily more expensive. However, as mentioned previously, no shortage is seen for aviation purposes until at least the year 2000 and then synthetic fuel is likely to be available. It is assumed, therefore, than an increasing number of people will fly for pleasure, take holidays, and so on, despite increasing costs.

At the same time, the microelectronic revolution also means that a lot of business can be conducted by television and telephone. The high value but low volume of electronic components means that air cargo is probably the best method of transporting such components! Already Heathrow airport carries more cargo than Liverpool and Southampton docks. The uses to which aeroplanes can be put seem inexhaustible and with rapidly developing technology there may be many more as yet not envisaged.

A few years ago it would have seemed absurd to shoot a rocket at the moon, especially one which carried people. Space travel is a natural extension of aeronautics, but, like most new fields of technology, it consumes very large amounts of money. Only two countries in the world have allocated the facilities to carry out this research in depth – the USA and the USSR. Even for these two wealthy nations the expenses have been so great that they have carried out joint missions in space.

A significant milestone in the American space programme was of course when the *Apollo 11* rocket (Fig. 11.27) took Neil Armstrong, Mike Collins and Buzz Aldrin into history. Whilst Mike Collins was orbiting the moon, in the Command Module, the Luna Module, named *Eagle*, had landed at the Sea of Tranquillity on 20 July 1969. Out stepped Neil Armstrong, then Buzz Aldrin, the first men from planet earth to set foot on its satellite, the moon. Later, on 15 July 1975, the joint USA/USSR mission had a *Soyuz 19* docking with an *Apollo*.

Fig. 11.27 *Apollo 11* blasts off on 16 July 1969

Contrast these events with the Wright Brothers flight at Kitty Hawk on 17 December 1903. In less than 70 years phenomenal progress has been made in the field of aeronautical engineering. When you consider more recent developments such as the space shuttle (Fig. 11.28), progress over the next 70 years is likely to be just as spectacular.

Fig. 11.28 The space shuttle prior to take-off

12 Design Characteristics

■ Introduction

In this chapter we consider the main requirements for several completely different designs. Any designer must remember that there is no single answer. The best answer is necessarily a compromise, as the best aircraft would probably cost too much to build or operate.

■ Short Haul Passenger and Mail Carrier

The aircraft for this purpose would be fairly small and would not fly at supersonic speeds. Therefore, the body could have a bulbous shape and the wing could be straight with little sweepback. Such a wing is inexpensive to construct. The wing must be positioned both to give stability and to give easy access to the fuselage for passengers and cargo and be capable of providing safe, low speeds for landing and take-off on short rough runways.

A propeller-driven aircraft would probably be best since the aircraft would be flying at a fairly low speed. A turbo-prop is more reliable than a piston engine, and uses less fuel and is much quieter than a pure jet engine.

■ Crop Sprayer

A single-seat plane of the hedge hopper type is what is needed. It could have either a high wing or dihedral wing, possibly with leading edge slots to give it stability at low speed – it could even be a biplane! The power system would probably need to be the low speed, piston-driven propeller variety such as a two or four cylinder air-cooled engine which is light and easily maintained. Much of the fuselage might be occupied with insecticide tanks, pumps and spray jets.

■ Short Range Mass Carrier

An aircraft like this would need to carry large numbers of passengers over relatively short distances. The aircraft fuselage would have to be wide to take this number of passengers or large containers of cargo. Turbo-fan engines would probably be used to give the aircraft a high altitude and high speed flight. A large fan jet, high by-pass ratio engine could best be used since many of these aircraft fly at night when high noise levels are especially noticeable and disturbing. This type of engine is also economic in its use of fuel and is widely used so that spare parts and qualified maintenance engineers and aircrew would all be readily available.

■ Fighter Bomber

This would need to have only one or two seats while the rest of the aircraft would be occupied by engine, fuel and armaments. A large engine of pure jet design would be needed to give high speed, acceleration and altitude. Fuel economy would not be a primary concern in this aircraft. Possibly vectored thrust, as in the *Harrier*, would be attractive, even though this tends to restrict the top speed. A swing wing would give a high top speed and a low speed at approach.

■ Aircraft to Rescue 'Downed' Crew

This could be a seaplane able to come down on and take off from the sea. However, rough seas would restrict its use in this way. An aircraft that can 'hover' in air would be a better bet and this can only be achieved by a helicopter. Moreover, a helicopter can be used for rescue in other circumstances, e.g. mountain tops and cliff edges, to which a seaplane would have absolutely no access!

■ Concorde – A Case Study

The design of a complete aircraft is a very complex business. Difficulties can occur which were not expected and solutions have to be found quickly. It is extremely expensive to stop the production schedule while the designers grapple with an unforeseen problem.

In building *Concorde*, the designers faced problems which had never been tackled before – how to build a passenger aircraft which could fly at more than twice the speed of sound, be as safe as possible and still be able to be flown economically. This section outlines just two of the specific difficulties which arose during the building of *Concorde*.

Moving the Centre of Gravity

THE PROBLEM
As we saw in Chapter 5, for an aircraft to fly there has to be a balance between the centre of pressure on the plane and its centre of gravity. The problem for the *Concorde* designers was that at supersonic speeds the centre of pressure moves *forwards* in the plane. Such a change would cause the plane to pitch unless the centre of gravity could be made to shift *backwards*.

At low speeds, the centre of gravity can be adjusted by using trim tabs. However, this causes drag and at supersonic speeds the drag produced would lead to a considerable increase in fuel consumption, making *Concorde* too expensive for commercial use.

THE SOLUTION

To solve the problem, the designers built a special fuel system. As the plane approaches supersonic speeds, the pilot operates servo mechanisms which cause some of the fuel in the front tanks to be pumped into tanks further back in the fuselage. This operation shifts the centre of gravity to the point where it is in balance with the centre of pressure. These ideas were tested on a full-scale test rig of the fuel system before they were tried out in the air.

The Choice of Materials

THE PROBLEM

Concorde was designed to cruise at Mach 2.2, but at such high speeds the aircraft produces a lot of friction in passing through the air. The friction causes the plane's skin to become very hot – an effect known as **kinetic heating**. Some parts of *Concorde* heat up to 150°C during supersonic flight. This temperature is close to the upper limit for aluminium alloys. Also, during descent, the temperatures can drop to -20°C in about 15 minutes. So the designers had to contend with both very high and very low temperatures, and very rapid changes.

The problem which the designers faced was to choose and design materials for the aircraft skin which could withstand such repeated exposure to heating and cooling during flight.

THE SOLUTION

First the most suitable alloys were selected and a series of test models was built. The small models were heated to the appropriate temperatures and then examined for defects. When this first stage had been completed satisfactorily, it was necessary to repeat the testing on a full-size model. Such tests were very difficult to carry out. Figure 12.1 on page 123 shows the range of temperatures which exist on the plane as it cruises at Mach 2.2.

The designers had to find a way to heat up different parts of the aircraft skin to their own particular temperatures. Eventually, a special kind of 'hot-water bottle' heating system was developed. The test plane was lagged with a water jacket, and superheated water was used to carry out the heating.

The engineers found that after 18 000 simulated flights cracks appeared in the wings. New designs had to be made to eliminate the source of the cracking. In fact, during the development of *Concorde* about four different versions were produced. Because the testing was so thorough and

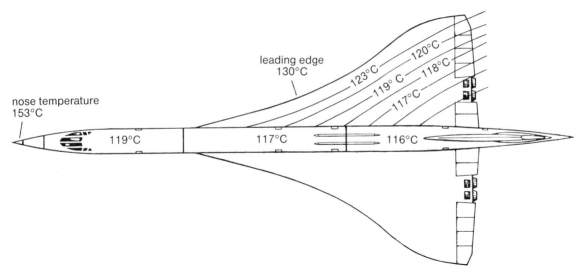

Fig. 12.1 *Concorde* skin temperatures at Mach 2.2

complex, *Concorde* is probably one of the best designed and safest aircraft in the world.

■ Conclusion

From everything we have looked at, you will see there is rarely, if ever, one unique way of doing something well.

This must mean that in some aspects of design, there are still exciting improvements to come. As new materials are developed and as new priorites emerge over the relative cost and values of materials, fuels, time, comfort and last but not least, safety in aeronautics, we can expect new designs at ever increasing speeds, and even more efficient aircraft.

Index

Note: illustrations are shown in bold numbers. In most cases, there are also textual references on these pages.

acceleration 32, 68, 77
accidents
   aeroplane 11, 37–8, 107, 111, 115
   airship 9, 10, 16
   gliding 22, 24
   jet plane 11
   see also stalling
action/reaction forces 50–**1**, **55–6**, **84**
   see also Newton's third law
advertising 13, 17
aerial photography 8, 17, 102
aerials, radio 18
aerofoils 41, 84
   forerunners 8
   see also slats *and* wings
aeroplanes
   accidents *see under* accidents
   care of 47
   construction **32–9**
   control *see under* control
   delta-plan 111
   delta wing 79 (*see also* delta wings)
   design *see* design
   engines **58–71**
   fabric covering 32, 33, **34**, 36
   fighter 72, 76, 89, 102–3, **106**, **109**, 121
   gliding by *see under* gliding
   history 11, **23–4**, 101–6, 109–15, 119
   jet *see* jet planes
   materials 32–3
   power units 55–71
   pressurised 36–7, 38
   principles 40–54, 56
   reactions to 100
   service problems 100–1
   skins 33, **37**, 38
   speed *see under* speed
   stability *see* stability
   structure **31–9**
   turbulence and 21
   vertical take-off 12, 114
   wind speeds and 21
   wings *see* wings
afterburning **70**
agility 102
agricultural aircraft **115**, 120

ailerons 78, 79, 80, **81–2**, 96–7
air 14, 15
   as coolant 59
   oxygen content 36–7
   pressure 36–7, **40**–1, 60
      flight instruments dependent on 88–**9**
   speed 41
   temperatures 64
air balloons *see* hot-air balloons
air corridors **98**–9
air-sea rescue **113**
air traffic control 97–**9**, 109
air traffic control transponders 99
aircraft *see* aeroplanes, airships *and* helicopters
aircraft control authorities 39
aircraft industry 12
airflow 41, **47–8**; *see also* turbulence
airports 97, 98, 99
airscrews *see* propellers
airships **9–10**, 15–17
   accidents 9, 10, 16
   uses 9, 10
airspace, controlled **98–9**
airspeed indicator 87–**8**
Akron airship 9
Alcock, John 103
Aldrin, Buzz 118
alloys 35, 39, 104, 122
altimeter 87–**8**
altitude 61, 87, 91, 99
aluminium 32, 39
   alloys 35, 39, 104, 122
America *see* USA
angle(s)
   coning **85**
   drag **85**
   flapping **85**
   gliding 26, **28–9**
   of attack 24, **44–5**, 54, 56
   pitch 57–8
   sweep **94–5**
annulus chamber 64
Apollo 11 **118**
area and drag 49–50
area ruling 50
Argonaut **110**
Armstrong, Neil 118
aspect ratio **30**, 49, 84
athletes 55
atmospheric pressure *see under* air
attack, angle of 24, **44–5**, 54
   propellers' 56
Auster 33, **34**

autopilot 88
Avro York 110
axes 17, **18**

BAC1-11 68
balance 45
   in kites 17, **18**
   passengers and 52, 83
balance of forces *see under* forces
ballast 13, 16
balloons and ballooning **8**–9, **13–15**, 107
   see also airships
banking 24, 80, 81–2, 83; *see also* rolling
barrage balloons **13**
BE2 102
Bell X-1 71
Bell X-2 71
Bernoulli's principle 40–2, 44
biplanes 20, 35, 36, **104**
   and drag 46
birds 24, 31, 58
   flying machines copying 10
Bleriot, Louis 102
BOAC 11, 111
BOAC *Argonaut* 110
Boeing 36
Boeing 247D 105
Boeing 307 36
Boeing 707 111
Boeing 747 38, 111–**2**
bombers 9, 110, 121
bombs 9, 102–3
bones 31
boss (of propeller) 57
bow wave 49
bowed kites **20**
box kites **18–19**
brake effect 83; *see also* speed, reducing
Britain 11, 12
   Civil Aviation Authority 39, 99
   flight information regions 97
   use of airships 16, 17
Britannia 66
Britten-Norman *Islander* **116**
Brown, Arthur Whitten 103
buildings and turbulence 21
bungee tow **25**
by-pass jet engine *see* turbo-fan engine

CAA 39, 99
camber 82, **83**, 94
Camel 102

124

Canadair *CL-215* 116
cantilevers 35, **104**
carburettors 41–2
cathode ray tubes 99
Cayley, George 8, 10–11, 22, 58
centre of gravity 43–4, **53**, 73
    and spin 52
    and yawing **77**
    changes in 52, 121–2
centre of pressure 43–4, 45, 52, 121
centrifugal force **85**
ceramic foam tiles 39
Charles, Jacques 8
China 7, 10
chord line **44**, **45**
circuses, flying 107
Civil Aviation Authority 39, 99
CL-215 **116**
clubs 104, 107
Cockerell, Christopher 43
coded signals 99
collective pitch change 84, 86
Collins, Mike 118
colours 47
combustor (in jet engine) 64–5
Comet I 11, 37–8, **111**
Comet IV 38, 111
communications 104, 116
    kites used for 7, 18
compasses 87, 88, 107
    gyroscope 88, 89
competition 11
competitions 101–2, 106
compression wave *see* shock wave
compressors **60–1**, 64–5, **67–8**
computer control 89–90
Comu 11
Concorde 12, 38, 89, 97, 100, 115, **117**, 121–**3**
    engine 68–**9**
coning angle **85**
connecting rod **59**
control
    air traffic *see* air traffic control
    of aeroplanes 24, 35, 54, 78–**83**, 102, 114
    (delta-winged) 96–**7**
    of balloons 8
    of helicopters 84, 86
    of kites **19**, 20
control authorities 39
control column (joystick) 78, 79, **81**
control flaps *see* ailerons
control zones **98**
controlled airspace **98**–9
coolants 59
cooling 59
corridors, air **98**–9
cracks 38, 122
crankshaft **59**
crop spraying 115, 120
currents, thermal **26**, 29

curvature, wing 22, 82, **83**, 94
cyclic load conditions 38
cyclic pitch 84, 86

da Vinci, Leonardo *see* Leonardo da Vinci
Daily Mail prize 101–2
Dakota **105**
DC2 105
DC3 Dakota **105**
DC4 110
DC8 111
DC10 111
De Havilland 11, 37; *see also Comet* and *Moth*
delta wings 27, 79, **96–7**
design (aeroplane) 12, 30, 38, 39, 100–1, 111, 120–**3**
    and speed 92
    and stall-avoidance 45
    testing 24, 29, 39, 111, 122–3
diffusion 15
dihedral (on kites) 20
dihedral wings 74, **75**–6
direction indicator 87–**8**
disturbance 41, **92**–4; *see also* turbulence
doping 33
Douglas DC see DC
Douglas Skystreak 92
drag **28–9**, 45–50
    balloons and 8
    flaps and 82, **83**
    form 46
    friction and 46–7
    gliding and 26, 28–9
    induced 30, 47–9
    interference 49–50
    kites and 17, **18**, **19**, **20**
    shape and 8
    shock wave and 49, **93**–5
    speed and 45, 47, 49
    spoilers and 83
    surfaces and 29
    thrust and 51–2
    types 45, 49
    wings and 30, 49–50
drag angle **85**
drag hinge 85
duralumin 35, 39

E66 92
E166 92
Earhart, Amelia 107
echos 99
Eckener, Hugo 9
elevators 54, 78, **79**, 80
    forward 24
employment 12
energy *see* power
engines 41–2
    aeroplane 11, 24, **58–71**, 120
    efficiency compared 67, 68
    airship 9

car 59
    helicopter 11–12
    hovercraft 42
    jet *see* jet engines
equilibrium 73
Europa airship 17
exports 12

F86 92
F100 92
F104 92
F111 **95**–6, 114
Fairey Delta 2 92
fatigue 38
feathering 57–8
feminists 107
fin *see* tailfin
firefighting 116
fishery protection 17
flapping angle **85**
flapping hinge 85
flaps 82, **83**
flexible-wing kites 26–**7**
flight
    history 7–12, 101–15, 119
    human-powered 10, 58, **101**
    uses 8, 10, 12, 17, 100, 102–3, 113, 115–16
    and design 120–1
    see also aeroplanes, airships *and* helicopters
flight information regions 97
flight instruments 8, 87–90
floaters (for kites) **19**
fluids 14
Flyer 1 **23**
Flyer 3 101
flying boats **105**
flying circuses 107
foam ceramic tiles 39
Focke 11
Fokker triplane 35
force(s) **28–9**
    and acceleration 68, 77
    and control 82
    and kites 17–**18**
    and stability 73–8
    balance of 50–**4**
    centrifugal **85**
    couples 44, 52
    see also action/reaction forces *and* drag
form drag 46
forward elevators 24
France 8, 16
Franklin, Benjamin 8
friction 46–7; *see also* drag
fuel(s) 64, 100, 118
    consumption 63, 70, 100
    moving system 122
fuselage construction 32–**3**

gas-filled balloons 8, 9; *see also* airships

125

gases 14, 15; *see also* helium *and* hydrogen
gearboxes 67
geodetic structure **36**
Germany
   development of gliders in 11, 25, 26
   development of helicopter in 11–12
   development of jet plane in 11
   USA and 9–10
   use of airships 9–10, 16
Gigant 26
gliders 8, **10–11**, **22–30**
   pilot's position 29
   principles of **26, 28–30**
   tows for 26
gliding
   accidents 22, 24
   by aeroplanes 28, 44, 53
   by space shuttle 89
   sport 30, 117
   theory of **26, 28–30**
gliding angles 26, **28–9**
Gloster Gladiators 35
Gloster Meteor 92, **109**
Goodyear 17
Gossamer Albatross **101**
Gossamer Condor 101
Graf Zeppelin airship **9**
gravity 32
   centre of *see* centre of gravity
Great Britain *see* Britain
ground control 97–**9**
gyroscope compasses 88, 89

Halifax bomber 110
Halton 110
hang-gliders and hang-gliding 11, **22**, 26–**7**, 30, 117
Hargraves, Lawrence 8
Harrier **114**, 121
Hawker Hunter 92
Hawker Hurricane 36
heat insulation 39
Heath, Lady 107
heating 122
height 61, 87, 91, 99
helicopters 11–12, **84–6**
   uses 12, 113, 121
helium 9, 14
   in airships 16, 17
   in balloons 14
helix **57**
high wings 74–**5**
hills
   and gliding 26
   and wind 21
Hindenburg airship 9, 10, **16**
hinge, drag 85
hinge, flapping 85
horizon, artificial 87–**8**
hot-air balloons **8**, 13, 14–**15**
HOTOL 71

hovercraft 12, 42–**3**
hovering 13, 14
human-powered flight 10, 58, **101**
Hurricane 109
hydrogen 9, 14
   in airships 9, 15–16
   in balloons 9, **14**

induced drag 47–9
inertial guidance 89
information 99; *see also* flight instruments
information regions, flight 97
inspection procedures 38
instruments, flight *see* flight instruments
interference drag 49–50
Islander **116**

Japan 7
jet engines 55, 56, 61–**2**, **64–70**
   compared to propellor engines 63
jet planes 11, 37–8, **111**
jobs 12
Johnson, Amy 107–**8**
joystick *see* control column
Junkers 35

keel area 77
keel kites **20**
kites 7–8, **17–21**
   flexible-wing 26–**7**
   keel **20**
   safety precautions for 20
   uses 8
   whether real flight 10
Korean War 113
Kremer prizes 101, 102

L17 **104**
laminar flow 41
Lancaster bomber 110
Lancastrian **110**
landing 58, 83, 97, 114
lateral stability 74–7
Leonardo da Vinci 10, 11
lift 14, 51–**3**
   and aerofoils 41
   and angle of attack 24
   and balloons 14–15
   and controls **82**, **83**
   and gliding 26, 28
   and kites 17, **18**, 20
   and rolling **74–7**
   and speed 45
   in helicopters 84, **85**, **86**
   spoilers and 83
   wings and 24, 43–5, **76–7**
Lilienthal, Otto 11, 22, 24, 27
Lindenberg, Charles 87
load 13, 14
load conditions 31–2, 38
Lockheed F104 92

Lockheed P80 92
Lockheed Tristar 111–**12**
Lockheed YF12A 71, 92
longerons 32–**3**, 35
looping the loop 107

Mach, Ernst 91
Mach numbers/speeds 49, 61, 91, 92
   critical 94
machined sections 38–**9**
magnetism and magnetic fields 88
mail 104
man-carriers (kites) **7**
man-lifters (kites) **7**
man-powered flight 10, 58, **101**
manoeuvrability 35, 72, 76, 102; *see also* control
masks 36, 37
mass
   and acceleration 68, 77
   see also weight
materials 32–3, 35, 36, 39, 104–5, 122
   viscosity 47
Messerschmitt Me 110 26
Messerschmitt Me 163 71
Messerschmitt Me 262 109
Messerschmitt Me 321 26
metal fatigue 38
metals 39; *see also* aluminium
meteorology 8, 13
Mitchell, R. J. 106
moments 52, 53
monitoring *see* air traffic control
Monocoque Duperdessin 33
monoplanes 35–6, 46
Montgolfier brothers 8
Mosquito 36
Moth **104**
motion, planes of **72**
motors *see* engines
muscle power 101; *see also* human-powered flight

navigation 87–90, 107
Newton's second law 68, 77
Newton's third law 51, 55–6, 62, 84
noise 63, 100, 120–1
   reducing 68, 69

O'Berg, Mrs Hart 107
observation 102; *see also* photography
Ohain, Pabst von 11
Olympus 593 engine 68–**9**
ornithopters **10**, 58
oxygen 36–7, 71

P80 92
Panavia Tornado 114
parachutes and parachuting 26, 117
passenger services 103, 111

passengers 107, 120
   effect of movement 52, 83
petrol engines 11, 41–2
photography, aerial 8, 17, 102
Pilcher, Percy 22
pilots
   in aeroplanes 45 (*see also* control *and* navigation)
   in gliders 29
piston engines 65, 109
   compared to turbo-prop engines 66
pistons 59
pitch angle 57–8
   in helicopters 84, **86**
pitching and pitch plane 72, 73–4
   control of 78–9
Pitot head 88–9
planes *see* aeroplanes
planes of motion **72**
power 11, 58, 65
   and weight 11, 63
   see also engines
powered flight *see* aeroplanes
pressure 36–7, 38, **40**–1
   air *see under* air
   and induced drag 47–9
   centre of 43–4, 45, 52, 121
   flight instruments dependent on 88–**9**
   in jet engines 64, 65
pressurisation 36–7, 38
prizes 101–2
propane 14, 15
propellor engines 58–**61**, 109
   compared to jet engines **63**
   see also turbo-prop engines
propellors 8–9, 24, 55, **56**–**9**
   future possibilities 68
   path of blade 57
   variable pitch 57–8
pullers (for kites) **19**
pumps 60; *see also* compressors

Quimby, Harriet 107

R101 airship 9
races 106
radar 17, 97, 98, **99**, 109
radial engines **59**
radio 18, 97, 99
ram jet **70**
ratio(s)
   aspect **30**, 49, 84
   lift/drag 26
   power/weight 11, 63
   strength/weight *see under* strength
RB211 engine 65, 68
reaction forces 50–**1**, **55**–**6**, **84**; *see also* Newton's third law
reheating **70**
research 12; *see also* design
rescue **113**, 121

resistance and propellor pitch 57–8
reversed pitch 58
Richey, Helen 108
Richthofen, Manfred von 35
rigidity 31, 33, 35
   and size 17
ripples **92–3**
roar 63
rockets 70–**1**, **118**
Rogallo, Francis 26
rolling and roll plane **72**, 74–7, 96–7,
   and yawing 77, 80
   control of 78, 80, **81–2**
rotating wing *see* rotor
rotation **51–3**
rotor 64, **65**, **84–6**
rudder 24, **77**, 78, **80**, 82
rudder pedals 78, 80
Rutan Long EZ **117**

safety 31–2, 38, 39, 99, 123
   in kite- flying 20
Scotland 7
screws 57
SE5 102–**3**
Sea King helicopter **113**
shock stall 94
shock wave 49, **93**–5
   reducing effect 95–6
Short brothers 105
Short flying boats **105**
Short *L17* **104**
shroud lines 26
sideslip 74, 75–6; *see also* rolling
signals 99
Sikorsky, Igor Alexis 11, 12
simulations 111, 122
size and rigidity 17
skin 33, **37**, 38
skin friction 46–7; *see also* drag
Skyship 500 airship 17
Skyship 600 airship 17
slats **83**
sled kites **20**
slipstream 56
slots **83**
sonic flight 91–5
Sopwith Camel 102
sound
   speed of 49, 61, 91, 93
   see also noise
sound barrier 91–2
sound waves **93**
Soyuz 19 118
space shuttle 39, 71, 89–**90**, **119**
space travel 118
speed 38, 91–2
   aeroplanes' 11, 66, 98, 102, 106
   and engine type 63, 67
   and variable geometry wings 114
   combined 97
   landing 58

air 41
airships' 9
and drag 45, 47, 49
and fatigue 38
and lift 45
and temperature 39
limitations on 61
measuring 91
of hovercraft 43
of propellor blades 61
of sound 49, 61, 91, 93
reducing 82, 83
see also supersonic flight *and* velocity
speed indicators 87–**8**
spin **51**–2
Spitfire **106**, 109
spoilers **82**–3
sport 13, 30, 104, 107, **117**
SRN1 **43**
SRN2 **43**
stabilisers *see* tail planes
stability 20, 53–**4**, 72–8, 102; *see also* control
stalling 17, 22, 28, 41, 44–5, 53
   prevention 83
   shock 94
stators **64**
steel 39
stiffness *see* rigidity
Stratoliner 36
streamlining **46**
strength 31, 51
   and weight 20, 23, 31, 36
stressed skin **37**, 38
stringers 33
struts **27**, 32–**3**
stunter kites **19**
stunts 107
submarines 13
subsonic flight 91, 92–3
super-charger compressors **60**
Supermarine S6B **106**
Supermarine Spitfire **106**, 109
supersonic speeds and flight 38, 68, 90–7, 114
   and temperature 39
   engines for 68–**9**
   navigation in 89
surfaces 29
swashplate 86
sweep angle 94–5
swept-back wings 74, **76**–7, **95**–6, 111, 114
swing-wing aeroplanes 114
Swordfishes 35

tailfin 54, **77**, **80**
tail planes 53–4, 73, **79**
tails (for kites) **19**, 20
take-off 12, 83, 97, 114
   vertical 12
teamwork 12, 5, 30

127

temperatures 64, 65, 122–**3**
   and engine efficiency 67
   and speed 39
thermals (thermal currents) **26**, 29
thrust 51–2, **53**
   and drag 51–2
   and engine type 63
   as reaction 56
   jet engines and 62, 65
   reheating and 70
   reversing 58
Tiger Moth 33
titanium 39
transonic flight 91–5
transponders, air traffic control 99
transport 116, 118
trees and turbulence 21
Trident 68, 79
trim 45
trim tabs 83
triplanes 35
Tristar 111–**12**
Tupolev Tu 114 66
turbines 65
turbo-charger compressors 60–1
turbo-fan engines 65, **67**–8, 69
   uses 120
turbo-jet engines **62**, **64**–5, 109
   limitations 67, 68
turbo-prop engines 65, **66**–7
   uses 120
turbulence **19**, 21, 41, 44
   and form 46
   and kites 21
   shock wave and **93**–4
turn and slip indicator 87–**8**
turning **80**–2
   measuring 87
twin-spool engine **66**

UK *see* Britain
USA 118
   and Germany 9–10
   early kite experiment in 8
   use of airships 10, 16, 17
USSR 118

V-1 missiles 70
variable geometry wings 114
VC10 **111**

velocity 68; *see also* speed
Venturi gauge **40**, 41
Vertical Short Take Off and Landing 114
vertical speed indicator 87–**8**
vertical take-off aircraft 12, 114
Vickers *Supermarine S6B* **106**
Vickers *Supermarine Spitfire* **106**, 109
Vickers *Swift* 92
Vickers *VC10* **111**
Vickers *Vimy* **103**
Vinci, Leonardo da *see* Leonardo da Vinci
viscosity 47
Viscount 66
visibility 87, 88, 98, 107
vortices 30, 43, 47
V/STOL 114

waisting 50
Wallis, Barnes 96
war 11; *see also* Korean War, World War I *and* World War II
water
   as ballast 16
   as coolant 59
waves 49, **92**–**3**; *see also* shock wave
weather 98, 107; *see also* meteorology
weight 13–15, **28**–**9**, 31
   and acceleration 32
   and power 11, 63
   and strength 20, 23, 31, 36
   in dynamic situations 51
   see also mass
Wellington **36**
Westland *Sea King* helicopter **113**
Westland *Whirlwing* helicopter **113**
whirlpools 30, 43, 47
Whirlwing helicopter **113**
Whittle, Frank 11, 109
Wilson, Alexander 7–8
wind 21
   and balloons 8
   and flying direction 87
   and kites 19, **21**
   speed 45
wind tunnels 24, 29, 47
wing walkers 107

wings (aeroplane) 17, **30**, 104
   and control 35
   and drag 30, 47, 49–50
   and lateral stability 74–7
   and lift 43–5, 82, **83**
   and shock wave **94**
   aspect ratio **30**, 49
   curvature/camber 22, 82, **83**, 94
   delta **96**–**7**
   dihedral 74, **75**–6
   fixed versus moving 22, 58
   flexible **26**–**7**
   high 74–**5**
   mean chord width 30
   shape 49, 82, **83**
   structure 35
   swept-back 74, **76**–7, **95**–6, 111, 114
   swing 114
   warping 2*d*
wires 32–**3**
women 107–8
wood 32
World War I 9, 16, 102
   aeroplanes used in 35, 102–3
   gliders used in **25**
World War II 13, 16
   aeroplanes used in 35, 36, 37, 47, 71, **109**
   gliders used in 25–6
   kites used in 18
   missiles used in 70
   women's employment in 108
Wright, Orville 11, 23–4, 58, 101, 119
Wright, Wilbur 11, 23–4, 58, 101, 107, 119

X-1 71
X-2 71

yawing and yaw plane 54, **72**, 77, **96**–**7**
   and rolling 77–**8**
   control of 78, **80**
YF12A 71, 92

Zeppelin, Ferdinand von 9
Zeppelin airships **9**
zones, control **98**